TYPE SPECIMENS OF RECENT MAMMALS IN THE MUSEUM OF VERTEBRATE ZOOLOGY, UNIVERSITY OF CALIFORNIA, BERKELEY

Type Specimens of Recent Mammals in the Museum of Vertebrate Zoology, University of California, Berkeley

by Blair Csuti

A Contribution from the Museum of Vertebrate
Zoology of the University of California, Berkeley

UNIVERSITY OF CALIFORNIA PRESS
Berkeley · Los Angeles · London

UNIVERSITY OF CALIFORNIA PUBLICATIONS IN ZOOLOGY

Editorial Board: Cadet H. Hand; George L. Hunt Jr.;
Peter Moyle; James L. Patton

Volume 114

Issue Date: December 1980

UNIVERSITY OF CALIFORNIA PRESS
BERKELEY AND LOS ANGELES, CALIFORNIA

UNIVERSITY OF CALIFORNIA PRESS, LTD.
LONDON, ENGLAND

ISBN 0-520-09622-3
LIBRARY OF CONGRESS CATALOG CARD NUMBER: 80-13264

©1980 BY THE REGENTS OF THE UNIVERSITY OF CALIFORNIA
PRINTED IN THE UNITED STATES OF AMERICA

Library of Congress in Publication Data

California. University. California Museum of Vertebrate
 Zoology.
  Type specimens of recent mammals in the Museum of
Vertebrate Zoology, University of California, Berkeley.

  (University of California publications in zoology;
v.114) (A Contribution from the Museum of Vertebrate
Zoology of the University of California, Berkeley)
  Bibliography: p.
  1. Mammals—Catalogs and collections. 2. Type
specimens (Natural History) 3. California. University.
California Museum of Vertebrate Zoology— Catalogs.
I. Csuti, Blair A. II. Title. III. Series: California.
University. University of California publications in
zoology; v. 114. IV. Series: California. University.
California Museum of Vertebrate Zoology. Contributions.
QL708.2.C34 1980   599'.0074.'019467   80-13264
ISBN 0-520-09622-3

Contents

Acknowledgments, vii

INTRODUCTION 1
LIST OF TYPES 3
GEOGRAPHIC ORIGINS OF TYPE SPECIMENS 65

Literature Cited, 75

Acknowledgments

I gratefully acknowledge the cooperation of the curators and staff of the Museum of Vertebrate Zoology throughout the course of this project. Much of the groundwork for this list was painstakingly and accurately laid by A. C. Ziegler during his service as a curatorial assistant under the direction of Seth B. Benson in the late 1950's. I am indebted to J. C. Hafner for valuable suggestions on both style and content, and to W. Z. Lidicker, Jr. and J. L. Patton for reviewing the manuscript. Impetus for the current project came from Sheila Kortlucke, whose notes and preliminary work on this list proved most instructive. I am grateful to Dianne Oshiro for carefully preparing the typeset copy of the manuscript.

INTRODUCTION

From its founding in 1908 until the advent of the Second World War, the mammal collections of the Museum of Vertebrate Zoology, University of California, Berkeley, enjoyed unprecedented growth, largely through the efforts of Joseph Grinnell, its first director, and Miss Annie M. Alexander, its founder and benefactor. By 1943 the collection exceeded 100,000 specimens, including over three hundred holotypes of Recent North American mammals. Miss Alexander herself collected 27 of these holotypes, the largest number contributed by a single individual. Donations of several larger private collections of mammals, some containing specimens previously described as holotypes, have further supplemented the museum's collection. Among these are parts of the collections of D. R. Dickey (primarily El Salvador material), O. P. Silliman, J. C. von Bloeker, and M. H. Swenk. Presently the collection numbers 156,000 specimens, and over the past ten years has grown at the rate of 1,800 specimens per year. Recent collection growth includes many specimens from Central and South America, reflecting current research interests at the museum.

Recognizing that the topographic and climatic diversity of the western United States provided a natural laboratory for the study of adaptation and evolution, Grinnell focused his efforts and those of his staff and students on this region. The resulting collections furthered our understanding of the nature of variation in animal species, and have provided the basis for numerous studies demonstrating the role of selection in the evolutionary process. A total of 334 holotypes of new species and subspecies are represented in the collections of the Museum of Vertebrate Zoology, and the current list represents the first compilation of these forms. Although 268 of these taxa were first described from localities in the western United States, many forms are included from South and Central America, as well as from central and eastern North America. A list of names by geographical origin follows the main body of the work.

In the following list, orders, families, and genera are arranged according to Simpson (1945). Species and subspecies are listed alphabetically by the name under which they were originally described. All types described below are holotypes. Following each name is the citation and actual date of publication of the original description. Age, sex, nature of the material (e.g., skin and skull, skull only),

Museum of Vertebrate Zoology (MVZ) mammal catalog number, and the type locality follow in order. Collector's information is taken from the original label. Supplementary information, often present in the original description, especially concerning the precise type locality, is given in brackets. The name(s) of the collector(s), date of collection, original number, and measurements appearing on the original specimen label follow the type locality. In instances where the collector was not the field preparator and cataloger, the latter's name is enclosed in parentheses. In so far as was possible, the current status of each taxon was investigated. Where the currently applicable name differs from that in the original description, the current name and citation of the latest revision are listed under "Remarks." Other pertinent information, such as comments on specimen condition, information from the collector's field notes, or published comments about the specimen, is included here as well. In most cases field notes accompanying the specimen are available, and these are housed in a separate collection at the museum.

Since animal taxonomy is a dynamic science, opinion will inevitably be divided over the status of certain names. The taxonomic status of two genera, *Spermophilus* and *Odocoileus*, remains contested (J. Mamm., 41:537-539, 1960). While recognizing that valid arguments may be advanced in favor of the use of *Citellus* as the genus for ground squirrels and *Dama* as the generic name for black-tailed deer, I have elected to use the names *Spermophilus* and *Odocoileus* in the present work, both of which have been generally adopted by North American mammalogists.

LIST OF TYPES

INSECTIVORA

Soricidae

Sorex arcticus maritimensis Smith

J. Mamm., 20(2):244, 15 May 1939.

Type.— Adult male, skin and skull, MVZ 84479; from Wolfville [Kings County], Nova Scotia; collected by Ronald W. Smith on 29 November 1936, original number 1444; measurements: 107-42-14.

Sorex cinereus nigriculus Green

Univ. California Publ. Zool., 38(7):387, 9 June 1932.

Type.— Adult, sex in question, skin and skull, MVZ 51413; from ["alluvial tidewater marsh on the Tuckahoe River, east of"] Tuckahoe, [Cape May County], New Jersey; collected by Morris M. Green on 23 December 1930, original number 154; measurements: 101-43-12.

Sorex halicoetes Grinnell

Univ. California Publ. Zool.,10(9):183, 20 March 1913.

Type.— Young adult male, skin and skull, MVZ 3638; from [salt marsh near] Palo Alto [Santa Clara County], California; collected by J. Dixon on 6 May 1908, original number 218; measurements: 108-40-12.

Remarks.— *Sorex vagrans halicoetes* (Jackson, N. Amer. Fauna, 51:108, 24 July 1928). Most of left parietal missing.

Sorex jacksoni Hall and Gilmore

Univ. California Publ. Zool., 38(9):392, 17 September 1932.

Type.— Adult female, skin and skull, MVZ 51142a; from [2 miles east of North Cape], Sevoonga, St. Lawrence Island, Bering Sea, Alaska; collected by R. M. Gilmore on 27 June 1931, original number 1658; measurements: 104-35-12-4.

Remarks.— Due to an error in cataloging, a series of 10 numbers was used twice. The second series of 10, including this type, were given the suffix "a."

Sorex melanogenys Hall

J. Mamm., 13(3):260, 9 August 1932.

Type.— Adult male, skin and skull, MVZ 50247; from Marijilda Canyon, Graham Mountains [=Pinaleno Mountains], 8600 feet altitude, Graham County, Arizona; collected by A. M. Alexander on 3 July 1931, original number 924; measurements: 107-42-13.

Remarks.— A synonym of *Sorex vagrans monticola* according to Findley (Univ. Kansas Publ., Mus. Nat. Hist., 9:50, 10 December 1955).

Sorex montereyensis mariposae Grinnell

Univ. California Publ. Zool., 10(9):189, 20 March 1913.

Type.— Young adult female, skin and skull, MVZ 12979; from Yosemite Valley, 4000 feet altitude, Mariposa County, California; collected by J. and H. W. Grinnell on 27 May 1911, original number (J. Grinnell) 673; measurements: 121-51-14.

Remarks.— *Sorex trowbridgii mariposae* (Grinnell, Univ. California Publ. Zool., 21:314, 27 January 1923).

Sorex obscurus malitiosus Jackson

Proc. Biol. Soc. Washington, 32:23, 11 April 1919.

Type.— Adult female, skin and skull, MVZ 8401; from east side of Warren Island [northwest of Prince of Wales Island], Alaska; collected by H. S. Swarth on 21 May 1909, original number 7532; measurements: 120-56-15.

Remarks.— *S[orex] m[onticolus] malitiosus* (Hennings and Hoffmann, Univ. Kansas Occ. Pap., Mus. Nat. Hist., 68:14, 15 July 1977). Type in fresh summer pelage.

Sorex obscurus mixtus Hall

Amer. Natur., 72(742):462, 10 September 1938.

Type.— Adult male, skin and skull, MVZ 70376; from Vanada, Texada Island [Georgia Strait], British Columbia; collected by R. A. Cumming on 4 May 1936, original number 1501; measurements: 110-48-12.

Remarks.— *S[orex] m[onticolus] mixtus* (Hennings and Hoffmann, Univ. Kansas Occ. Pap., Mus. Nat. Hist., 68:14, 15 July 1977). Moderately worn teeth (Hall, 1938).

Sorex ornatus relictus Grinnell

Univ. California Publ. Zool., 38(8):389, 9 June 1932.

Type.— Adult female, skin and skull, MVZ 51414; from [excavated slough immediately outside of] east side levee, Buena Vista Lake, 298 feet altitude, Kern County, California; collected by W. C. Russell on 26 February 1932, original number 1992; measurements: 98-39-12-7, 5.3 g.

Remarks.— Original description indicates altitude is 290 feet.

Sorex ornatus salarius von Bloeker

Proc. Biol. Soc. Washington, 52:94, 5 June 1939.

Type.— Adult female, skin and skull, MVZ 81548; from [saltmarsh at the] mouth, Salinas River, Monterey County, California; collected by Jack C. von Bloeker, Jr. on 13 August 1937, original number 8504; measurements: 95-33-12-5.

Sorex ornatus salicornicus von Bloeker

Proc. Biol. Soc. Washington, 45:131, 9 September 1932.

Type.— Adult male, skin and skull, MVZ 74679; from 1 mile north of La Playa del Rey, 25 feet altitude, Los Angeles County, California; collected by Jack C. von Bloeker, Jr. on 13 March 1932, original number 1680; measurements: 102-38-11-3.

Remarks.— Type described while in collection of Jack C. von Bloeker, Jr.

Sorex pacificus sonomae Jackson

J. Mamm., 2(3):162, 19 August 1921.

Type.— Adult female, skin with complete skeleton, MVZ 19658; from Sonoma County side of Gualala River, by Gualala [in Mendocino County], Sonoma County, California; collected by Alfred C. Shelton on 2 July 1913, original number 227; measurements: 133-59-16.

Remarks.— Regarded as a subspecies of *Sorex vagrans* by Findley (Univ. Kansas Publ., Mus. Nat. Hist., 9:32, 10 December 1955) but considered valid by Hennings and Hoffmann (Univ. Kansas, Occ. Pap. Mus. Nat. Hist., 68:18, 15 July 1977). The original description indicates the specimen is a male (teeth moderately worn), but the skin tag and collector's catalog indicate the specimen is a female, while the sex symbol on the original skull tag is ambiguous. Miller and Kellogg (1955:25) incorrectly indicate that the type specimen is in the United States National Museum by the presence of a dagger following the author's name.

Sorex sinuosus Grinnell

Univ. California Publ. Zool., 10(9):187, 20 March 1913.

Type.— Young adult female, skin and skull, MVZ 16470; from Grizzly Island [near Suisun], Solano County, California; collected by A. M. Alexander on 5 January 1912, original number 1902; measurements: 99-37-12.

Blarina brevicauda pallida Smith

Amer. Midland Natur., 24(1):223, 31 July 1940.

Type.— Adult male, skin and skull, MVZ 86682; from Wolfville [Kings County], Nova Scotia; collected by Ronald W. Smith on 5 October 1937, original number 1461; measurements: 121-38-14.5.

Talpidae

Neürotrichus gibbsii minor Dalquest and Burgner
Murrelet, 22(1):12, 30 April 1941.

Type.— Adult male, skin and skull, MVZ 94857; from [University of Washington campus] Seattle, King County, Washington; collected by Walter W. Dalquest on 19 May 1940, original number 1822; measurements: 107-35-15.

Scapanus latimanus campi Grinnell and Storer
Univ. California Publ. Zool., 17(1):1, 23 August 1916.

Type.— Adult male [in winter pelage], skin and skull, MVZ 21520; from Snelling, 250 feet altitude, Merced County, California; collected by C. L. Camp on 9 January 1915, original number 1746; measurements: 170-37-22, 86.6 g.

Scapanus latimanus caurinus Palmer
J. Mamm., 18(3):290, 14 August 1937.

Type.— Adult male, skin and skull, MVZ 25388; from Laytonville, Mendocino County, California; collected by F. C. Clarke on 14 May 1917, prepared by H. S. Swarth, original number (H. S. Swarth) 10590; measurements: 186-38-22, 97.4 g.

Scapanus latimanus grinnelli Jackson
Proc. Biol. Soc. Washington, 27:56, 20 March 1914.

Type.— Adult male, skin and skull, MVZ 17785; from Independence, 3900 feet altitude, Inyo County, California; collected by H. A. Carr on 8 May 1912, original number 915 (not given in original description); measurements: 156-31-21.

Remarks.— Grinnell (1933:77) gives the type locality as: "site of old Fort Independence (on ranch of Carl Walters), 2 miles north of Independence."

Scapanus latimanus insularis Palmer
J. Mamm., 18(3):297, 14 August 1937.

Type.— Adult male, skin and skull, MVZ 68993; from Angel Island, San Francisco Bay, Marin County, California; collected by F. G. Palmer on 20 October 1935, original number 767; measurements: 177-38-23.

Scapanus latimanus monoensis Grinnell
Univ. California Publ. Zool., 17(14):423, 25 April 1918.

Type.— Adult female, skin and skull, MVZ 25834; from Taylor's Ranch, 5300 feet altitude, 2 miles south of Benton Station, Mono County, California; collected by H. G. White on 29 August 1917, original number 1376; measurements: 151-32-20.

Scapanus latimanus occultus Grinnell and Swarth
Univ. California Publ. Zool., 10(3):131, 13 April 1912.

Type. — Young adult female, skin and skull, MVZ 2369; from Santa Ana Cañon, 400 feet altitude [Santa Ana Mountains], Orange County, California; collected by H. S. Swarth on 20 September 1908, original number 7051; measurements: 150-33-18.

Remarks. — Grinnell (1933:76) adds: "12 miles northeast of Santa Ana."

Scapanus latimanus parvus Palmer

J. Mamm., 18(3):300, 14 August 1937.

Type. — Adult male, skin and skull, MVZ 30343; from Alameda, [Alameda County], California; collected by M. P. Anderson on 1 June 1916, original number 37; measurements: 160-33-18, 33 g.

Remarks. — Miller and Kellogg (1955:46) give "Alameda Island" instead of "Alameda." The city of Alameda is located on Alameda Island.

Scapanus orarius yakimensis Dalquest and Scheffer

Murrelet, 25(2):27, 19 September 1944.

Type. — Subadult male, skin and skull, MVZ 96354; from 3/4 mile north of Union Gap, Yakima County, Washington; collected by J. A. Gray, Jr. on 3 July 1941, original number 1026; measurements: 162-26-20, 58 g.

Condylura cristata nigra Smith

Amer. Midland Natur., 24(1):218, 31 July 1940.

Type. — Adult female, skin and skull, MVZ 86603; from Wolfville, [Kings County], Nova Scotia; collected by R. W. Smith on 13 November 1937, original number 1513; measurements: 197-77-28.

CHIROPTERA

Vespertilionidae

Myotis californicus quercinus Grinnell

Univ. California Publ. Zool., 12(10):317, 4 December 1914.

Type. — Adult female, skin and skull, MVZ 6939; from Seven Oaks, 5000 feet altitude, San Bernardino Mountains [San Bernardino County], California; collected by J. Grinnell and J. Dixon on 8 July 1905, original number (J. Grinnell) 1120; measurements: 86-41-6.

Remarks. — A synonym of *Myotis californicus californicus* according to Miller and Allen (Bull. U.S. Nat. Mus., 144:152, 25 May 1928).

Myotis californicus stephensi Dalquest

Proc. Biol. Soc. Washington, 59:67, 11 March 1946.

Type. — Adult female, skin and skull, MVZ 16657; from Vallecito, San Diego County, California; collected by Frank Stephens on 29 March 1912, original number 3493; measurements: 81-39-7-12-221.

Remarks.— A renaming of *Myotis californicus pallidus* Stephens with a new type specimen. Right dentary, present when described, missing since 1958.

Myotis evotis pacificus Dalquest

Proc. Biol. Soc. Washington, 56:2, 25 February 1943.

Type.— Adult male, skin and skull, MVZ 94173; from 3-1/2 miles east and 5 miles north of Yacolt, 500 feet altitude, Clark County, Washington; collected by J. Chattin on 3 August 1940, original number 620; measurements: 90-45-7-19-tr. 9, 7 g.

Myotis lucifugus relictus Harris

J. Mamm., 55(3): 598, 20 August 1974.

Type.— Adult female, skin and skull, MVZ 47335; from Keeler, 3600 feet altitude, Inyo County, California; collected by Seth B. Benson on 30 July 1931, original number 1293; measurements: 86-37-10-16, 5.7 g.

Remarks.— Right wing affixed with wire and glue.

Myotis ruddi Silliman and von Bloeker

Proc. Biol. Soc. Washington, 51:167, 23 August 1938.

Type.— Adult male, skin and skull, MVZ 81549; from Lime-kiln Creek, 250 feet altitude [southwestern Santa Lucia Mountains], Monterey County, California; collected by Jack C. von Bloeker, Jr. [and R.L. Rudd] on 31 July 1937, original number (von Bloeker) 8385; measurements: 103-46-10-13-40-19-262.

Remarks.— A synonym of *Myotis volans longicrus* according to Benson (J. Mamm., 30:48, 14 February 1949).

Myotis yumanensis altipetens Grinnell

Univ. California Publ. Zool., 17(2):9, 23 August 1916.

Type.— Adult male, skin with complete skeleton, MVZ 23034; from 1 mile east of Merced Lake, 7500 feet altitude, Yosemite [National] Park [Mariposa County], California; collected by J. Grinnell on 19 August 1915, original number 3437; measurements: 88-36-9-12, 7.8 g.

Remarks.— A synonym of *Myotis lucifugus carissima* according to Miller and Allen (Bull. U.S. Nat. Mus., 144:50, 25 May 1928).

Myotis yumanensis lambi Benson

Proc. Biol. Soc. Washington, 60:45, 19 May 1947.

Type.— Old adult male, skin and skull, MVZ 38194; from San Ignacio [latitude 27° 17' N], 500 feet altitude [cape region], Baja California, Mexico; collected by Chester C. Lamb on 19 May 1927, original number 7685; measurements: 78-32-9-10, 4.1 g.

Myotis yumanensis oxalis Dalquest

Amer. Midland Natur., 38(1):228, 20 August 1947.

Type. — Adult female, skin and skull, MVZ 102011; from Oxalis [San Joaquin Valley], Fresno County, California; collected by Mary C. Ramage on 20 April 1945, original number 85; measurements: 88-34-10-13, tr. 8.

Myotis yumanensis sociabilis Grinnell

Univ. California Publ. Zool., 12(10):318, 4 December 1914.

Type. — Adult female, skin and skull, MVZ 5158; from [Old] Fort Tejon [Tehachapi Mountains, 3200 feet altitude], Kern County, California; collected by J. Grinnell on 23 July 1904, original number 715; measurements: 84-36-9.

Corynorhinus macrotis intermedius Grinnell

Univ. California Publ. Zool., 12(10):320, 4 December 1914.

Type. — Adult female, skin and skull, MVZ 7753; from Auburn [1300 feet altitude, Placer County], California; collected by Dr. J. C. Hawver on 31 July 1909, skinned by J. Grinnell, original number "b" (J. Grinnell 2387), measurements: 100-51-9.

Remarks. — A synonym of *Plecotis townsendii pallescens* according to Handley (Proc. U.S. Nat. Mus., 110:190, 1959). Right auditory capsule present in vial but broken off skull.

Molossidae

Eumops sonoriensis Benson

Proc. Biol. Soc. Washington, 60:133, 31 December 1947.

Type. — Old adult male, skin and skull, MVZ 82150; from Rancho de Costa Rica, 270± feet altitude, Rio Sonora, Sonora, Mexico; collected by Seth B. Benson on 2 May 1938, original number 5413; measurements: 185-67-18-30, 55.0 g.

Remarks. — *Eumops underwoodi sonoriensis* (Hall and Villa, Univ. Kansas Publ., Mus. Nat. Hist., 1:446, 27 December 1949).

LAGOMORPHA

Ochotonidae

Ochotona albatus Grinnell

Univ. California Publ. Zool., 10(2):125, 31 January 1912.

Type. — Adult female, skin and skull, MVZ 16223; from [near] Cottonwood Lakes, 11000 feet altitude, Sierra Nevada Mountains, Inyo County, California; collected by J. Grinnell on 3 September 1911, original number 1741; measurements: 183-15-29-28.

Remarks.— *Ochotona princeps albata* (Hall, Univ. Kansas Publ., Mus. Nat. Hist., 5:127, 15 December 1951). Frontal region of cranium broken out of skull and present as chips in vial.

Ochotona princeps clamosa Hall and Bowlus

Univ. California Publ. Zool., 42(6):335, 12 October 1938.

Type.— Adult male, skin and skull, MVZ 78100; from north rim of Copenhagen Basin, 8400 feet altitude [Bear River Range of southeast Idaho], Bear Lake County, Idaho; collected by W. B. Davis on 21 July 1937, original number 2645; measurements: 195-15-32-22.

Remarks.— Original label indicates specimen is from the Wasatch Mountains.

Ochotona princeps tutelata Hall

Proc. Biol. Soc. Washington, 47:103, 13 June 1934.

Type.— Adult male, skin and skull, MVZ 58519; from Greenmonster Canyon, 8150 feet altitude, Monitor Mountains, Nye County, Nevada; collected by W. C. Russell on 15 July 1933, original number 3101; measurements: 180-9 (tail broken)-29-22, 128 g.

Ochotona princeps utahensis Hall and Hayward

Great Basin Natur., 2(2):107, 30 June 1941.

Type.— Adult female, skin and skull, MVZ 95273, from 2 miles east of West Deer Lake, southeast Aquarius Plateau, Garfield County, Utah; collected by G. S. Cannon on 25 June 1938, original number 73; measurements: 200-30, 196 g.

Remarks.— The original description indicates the type locality is "2 miles west of Deer Lake, in Sec. 9, R 5 E, T 32 S, Salt Lake Meridian." However a letter from Hayward to Hall, dated 7 April 1941, indicates the locality is 2 miles east of West Deer Lake, as located on a Powell National Forest map. The original specimen label says merely, "Private Lake, S. E. Aquarius Plateau." Study skin very flat.

Ochotona schisticeps muiri Grinnell and Storer

Univ. California Publ. Zool., 17(1):6, 23 August 1916.

Type.— Adult male, skin and skull, MVZ 23480, from near Ten Lakes, 9300 feet altitude, Yosemite National Park [Tuolumne County], California; collected by Walter P. Taylor on 11 October 1915, original number 7720; measurements: 188-14-31-22, 145.5 g.

Remarks.— *Ochotona princeps muiri* (Hall, Proc. Biol. Soc. Washington, 47:103, 13 June 1934). Skin in winter pelage.

Ochotona schisticeps sheltoni Grinnell

Univ. California Publ. Zool., 17(14):429, 25 April 1918.

Type.— Adult male, skin and skull, MVZ 27560; from near Big Prospector Meadow, 11000 feet altitude, White Mountains, Mono County, California; collected by A. C. Shelton on 29 July 1917, original number 3414; measurements: 188-8-30-24, 132.5 g.

Remarks.— *Ochotona princeps sheltoni* (Hall, Mammals of Nevada, p. 593, Univ. California Press, 1 July 1946). Skin showing chiefly newly acquired winter pelage (Grinnell, 1918).

Ochotona taylori Grinnell

Proc. Biol. Soc. Washington, 25:129, 31 July 1912.

Type.— Adult male, skin and complete skeleton, MVZ 11292; from Warren Peak, 9000 feet altitude, Warner Mountains, Modoc County, California; collected by W. P. Taylor and H. C. Bryant on 18 July 1910, original number (W. P. Taylor) 3885; measurements: 180-8-27-23.

Remarks.— *Ochotona princeps taylori* (Hall, Univ. Kansas Publ., Mus. Nat. Hist., 5:133, 15 December 1951). Cranium damaged, left dentary, originally present, missing since 1958.

Leporidae

Lepus bairdii oregonus Orr

J. Mamm., 15(2):152, 15 May 1934.

Type.— Adult female, skin and skull, MVZ 54887; from 12 miles south of Canyon City, 5500 feet altitude, Grant County, Oregon; collected by Robert T. Orr on 25 June 1932, original number 557; measurements: 470-43-139-82, c. 42.

Remarks.— *Lepus americanus oregonus* (Dalquest, J. Mamm., 23:179, 14 May 1942). Left mandible broken.

Lepus californicus depressus Hall and Whitlow

Proc. Biol. Soc. Washington, 45:71, 2 April 1932.

Type.— Adult female, skin and skull, MVZ 47066; from City Creek, 1-1/2 miles south of Pocatello, Bannock County, Idaho; collected by W. B. Whitlow on 7 December 1930, original number 442; measurements: 490-80-120-120.

Remarks.— A synonym of *Lepus californicus deserticola* according to Davis (The Recent mammals of Idaho, p. 359, The Caxton Printers, Ltd., 5 April 1939). The original description erroneously gives the type locality as "1/2 mile south of Pocatello."

Lepus washingtonii tahoensis Orr

J. Mamm., 14(1):54, 14 February 1933.

Type.— Adult female, skin and skull, MVZ 38286; from 1/2 mile south of Tahoe Tavern, Lake Tahoe [Placer County], California; collected by J. Moffitt on 7 May 1927, original number M24; measurements: 383-25-129-101, 2 lbs 6 oz.

Remarks.— *Lepus americanus tahoensis* (Dalquest, J. Mamm., 23:176, 14 May 1942).

Sylvilagus bachmani macrorhinus Orr

Proc. Biol. Soc. Washington, 48:28, 6 February 1935.

Type.— Adult female, skin and skull, MVZ 51679; from Alpine Creek Ranch [3-1/2 miles south of and 2-1/3 miles east of Portola, 1700 feet altitude], San Mateo County, California; collected by E. L. Summer, Jr. on 18 April 1932, original number 138; measurements: 344-45-73-64.5, 780.7 g.

Sylvilagus bachmani mariposae Grinnell and Storer

Univ. California Publ. Zool., 17(1):7, 23 August 1916.

Type.— Adult male, skin and skull, MVZ 21867; from El Portal, 4000 feet altitude (*Adenostoma* association) on McCauley Trail, Mariposa County, California; collected by J. Grinnell on 7 December 1914, original number 2972; measurements: 326-30-73-80, 625 g.

Sylvilagus bachmani riparius Orr

Proc. Biol. Soc. Washington, 48:29, 6 February 1935.

Type.— Adult female, skin and skull, MVZ 57348; from [west side of the San Joaquin River], 2 miles northeast of Vernalis [in San Joaquin County], Stanislaus County, California; collected by Robert T. Orr on 11 November 1931, original number 448; measurements: 347-41-76-68.

Sylvilagus bachmani tehamae Orr

Proc. Biol. Soc. Washington, 48:27, 6 February 1935.

Type.— Adult male, skin and skull, MVZ 34971; from [Dale's, on] Paine Creek, 600 feet altitude, Tehama County, California; collected by J. and W. F. Grinnell on 26 December 1924, original number (J. Grinnell) 6183; measurements: 332-32-74-70, 537.5 g.

RODENTIA

Aplodontidae

Aplodontia chyrseola Kellogg

Univ. California Publ. Zool., 12(5):295, 15 April 1914.

Type.— Adult male, skin and complete skeleton, MVZ 13328; from Jackson Lake, 5900 feet altitude, Siskiyou County, California; collected by A. M. Alexander on 22 June 1911, original number 1441; measurements: 370-21-59.

Remarks.— A synonym of *Aplodontia rufa rufa* according to Taylor (Univ. California Publ. Zool., 17:454, 29 May 1918).

Aplodontia humboldtiana Taylor

Proc. Biol. Soc. Washington, 29:21, 24 February 1916.

Type.— Adult male, skin and skull, MVZ 21162; from Carlotta, Humboldt County, California; collected by H. E. Wilder on 4 January 1914, original number 1494; measurements: 365-35-58.

Remarks.— *Aplodontia rufa humboldtiana* (Taylor, Univ. California Publ. Zool., 17:470, 29 May 1918).

Aplodontia nigra Taylor

Univ. California Publ. Zool., 12(6):297, 15 April 1914.

Type.— Adult male, skin and complete skeleton, MVZ 20320; from Point Arena, Mendocino County, California; collected by C. L. Camp on 10 July 1913, original number 1003; measurements: 346-38-55-16.

Remarks.— *Aplodontia rufa nigra* (Taylor, Univ. California Publ. Zool., 17:479, 29 May 1918).

Aplodontia rufa grisea Taylor

Univ. California Publ. Zool., 12(16):497, 6 May 1916.

Type.— Adult female, skin and skull, MVZ 3751; from Renton [near Seattle, King County], Washington; collected by Frank Stephens on 4 October 1907, original number 294; measurements: 330-25-55-18.

Remarks.— A synonym of *Aplodontia rufa rufa* (Taylor, Univ. California Publ. Zool., 17:454, 29 May 1918).

Sciuridae

Sciurus hudsonicus picatus Swarth

J. Mamm., 2(2):92, 2 May 1921.

Type.— Adult male, skin and skull, MVZ 8767; from Kupreanof Island [25 miles south of Kake Village, at southern end of Keku Straits], Alaska; collected by H. S. Swarth on 23 April 1909, original number 7281; measurements: 317-125-51.

Remarks.— *Tamiasciurus hudsonicus picatus* (Anderson, Bull. Nat. Mus. Canada, Biol. Ser., 102:120, 24 January 1947).

Marmota caligata broweri Hall and Gilmore

Canadian Field Natur., 48(4):57, 2 April 1934.

Type.— Adult, unsexed, skin with complete skeleton, MVZ 51675; from Point Lay [arctic coast of] Alaska; 10 December 1931 (collected in early December, 1931; obtained from native resident by Charles D. Brower), no original number or measurements.

Remarks.— *Marmota broweri.* Considered specifically distinct by Rausch (Chromosoma, 16:618, 26 May 1965). Rausch (Arctic, 6:117, July 1953) indicates that since suitable habitat is lacking near Point Lay, the type specimen was probably taken during a hunting trip to the head of the Kukpowruk River, about 69° N latitude.

Marmota flaviventris fortirostris Grinnell

Univ. California Publ. Zool., 21(6):242, 7 November 1921.

Type.— Adult female, skin and skull, MVZ 27539; from McAfee Meadow, 11800 feet altitude, White Mountains, Mono County, California; collected by A. C. Shelton on 10 August 1917, original number 3533; measurements: 522-154-68-16.

Remarks.— Second label attached to the unstuffed skin indicates it was degreased in 1917 by J. Dixon--not tanned, no acid used, fleshed, soaked 12 hours in gasoline and drummed through sawdust.

Marmota ochracea Swarth

Univ. California Publ. Zool., 7(6):203, 18 February 1911.

Type.— Adult female, skin and partial skull, MVZ 5872; from [head of Fortymile Creek] Forty Mile [in Yukon Territory], Alaska; collected by C. L. Hall on 19 August 1901, original number 477; no measurements.

Remarks.— *Marmota monax ochracea* (Howell, N. Amer. Fauna, 37:34, 7 April 1915). Swarth (1911) indicates that the specimen was "prepared with the anterior portions of the skull left within the skin." These were subsequently removed, and the skull now consists of a fragment of the rostrum with left incisor attached and the lower jaw. An additional tag found on the skin gives the locality as "up the 40-mile River, in mountains, Forty-mile, Yukon Territory, Canada; on Alaska side." Skin very greasy.

Marmota vancouverensis Swarth

Univ. California Publ. Zool., 7(6):201, 18 February 1911.

Type.— Adult female, skin with complete skeleton; MVZ 12094; from Mt. Douglas, 4200 feet altitude, 20 miles south of Alberni, Vancouver Island, British Columbia; collected by E. Despard on 8 July 1910, original number 30; measurements: 720-240-105-30.

Remarks.— Rausch (Mammalia, 35:85, March 1971) confirms the distinctiveness of *Marmota vancouverensis.*

Marmota vigilis Heller

Univ. California Publ. Zool., 5(2):248, 18 February 1909.

Type.— Adult male, skin and skull, MVZ 418; from Coppermine Cove, west shore of Glacier Bay, Alaska; collected by J. Dixon on 14 July 1907, original number 69; measurements: 680-210-102.

Remarks.— *Marmota caligata vigilis* (Howell, N. Amer. Fauna, 37:61, 7 April 1915). The original description differs from the specimen label in that on the latter A. Hasselborg is listed as collector, with original number 69 AH. Specimen consists of skull plus poorly tanned (or untanned) flat skin.

Ammospermophilus nelsoni amplus Taylor

Univ. California Publ. Zool., 17(4):15, 3 October 1916.

Type. — Adult male, skin and skull, MVZ 16693; from [20 miles south of] Los Banos, Merced County, California; collected by R. H. Beck on 20 June 1912, original number (J. Grinnell) 1957; measurements: 260-75-43-7.

Remarks. — The species *Ammospermophilus nelsoni* is regarded as monotypic by A. H. Howell (N. Amer. Fauna, 56:183, 18 May 1938). Grinnell (1933:127) further defines the type locality as "near mouth of Little Panoche Creek, western Fresno County."

Callospermophilus chrysodeirus perpallidus Grinnell

Univ. California Publ. Zool., 17(14):429, 25 April 1918.

Type. — Adult male, skin and skull, MVZ 27488; from near Big Prospector Meadow, 10300 feet altitude, White Mountains, Mono County, California; collected by D. C. McLain on 26 July 1917, original number (J. Grinnell) 4334; measurements: 265-90-39-13, 177.5 g.

Remarks. — A synonym of *Spermophilus lateralis trepidus* (*Citellus lateralis trepidus* of Howell, N. Amer. Fauna, 56:206, 18 May 1938). The original description by Grinnell indicates that he was the collector, however his field notes confirm that he turned his trap line over to D. C. McLain on the occasion this specimen was captured.

Callospermophilus trepidus Taylor

Univ. California Publ. Zool., 5(6):283, 12 February 1910.

Type. — Adult male, skin and skull, MVZ 8240; from head of Big Creek, 8000 feet altitude, Pine Forest Mountains [Humboldt County], Nevada; collected by W. P. Taylor and C. H. Richardson, Jr. on 27 June 1909, original number 768; measurements: 268-90-44.

Remarks. — *Spermophilus lateralis trepidus* (*Citellus lateralis trepidus* of Howell, N. Amer. Fauna, 56:206, 18 May 1938).

Citellus beldingi crebrus Hall

Murrelet, 21(3):59, 16 December 1940.

Type. — Adult female, skin and skull, MVZ 70547; from Reese River Valley, 7 miles north of Austin, Lander County, Nevada; collected by W. B. Richardson on 18 May 1936, original number 1432; measurements: 282-70-43-14, 355.1 g.

Remarks. — *Spermophilus beldingi creber* (*Citellus beldingi creber* of Miller and Kellogg, 1955:194).

Citellus elegans aureus Davis

Recent mammals of Idaho, p. 177, The Caxton Printers, Ltd., 5 April 1939.

Type. — Subadult male, skin and skull, MVZ 71965; from Double Springs, 16 miles northeast of Dickey, Custer County, Idaho; collected by J. A. Donohoe on 21 July 1936, original number 71; measurements: 300-75-43-15.

Remarks. — *S[permophilus] elegans aureus* (Robinson and Hoffmann, Syst. Zool., 24:79, 12 June 1975). Original description erroneously states the type was collected on 20 July 1936.

Citellus grammurus tularosae Benson

Univ. California Publ. Zool., 38(5):335, 14 April 1932.

Type.— Adult male, skin and skull, MVZ 50935; from [the malpais at] French's Ranch, 5400 feet altitude, 12 miles northwest Carrizozo, Lincoln County, New Mexico; collected by Seth B. Benson on 28 October 1931, original number 1603; measurements: 488-206-60-21, 767 g.

Remarks.— *Spermophilus variegatus tularosae* (*Citellus variegatus tularosae* of Howell, N. Amer. Fauna, 56:145, 18 May 1936).

Citellus lyratus Hall and Gilmore

Univ. California Publ. Zool., 38(9):396, 17 September 1932.

Type.— Adult male, skin and skull, MVZ 51172; from Iviktook Lagoon [on coast about 15 miles west of Northeast Cape], St. Lawrence Island, Bering Sea, Alaska; collected by R. M. Gilmore on 7 July 1931, original number 1738; measurements: 381-114-54-7.

Remarks.— *S[permophilus] parryii lyratus* (Nadler, Vorontsov, Hoffman, Formichova, and Nadler, Comp. Biochem. Physiol., 44B:34, 15 January 1973). Howell (N. Amer. Fauna, 56:101, 18 May 1938) indicates that Gilmore defined the type locality as "about 35 miles northwest of Northeast Cape."

Citellus tereticaudus arizonae Grinnell

Proc. Biol. Soc. Washington, 31:105, 29 November 1918.

Type.— Adult female, skin and skull, MVZ 25485; from Tempe [1157 feet altitude], Maricopa County, Arizona; collected by H. S. Swarth on 30 May 1917, original number 10596; measurements: 243-76-35.

Remarks.— A synonym of *Spermophilus tereticaudus neglectus* (*Citellus tereticaudus neglectus* of Howell, N. Amer. Fauna, 56:187, 18 May 1938).

Spermophilus tridecemlineatus blanca Armstrong

J. Mamm., 52(3):533, 26 August 1971.

Type.— Adult female, skin and skull, MVZ 60426; from the Conejos River, 5 miles west Antonito, Conejos County, Colorado; collected by L. Kellogg on 16 August 1933, original number 2409; measurements: 224-80-30-4, 79 g.

Eutamias amoenus celeris Hall and Johnson

Proc. Biol. Soc. Washington, 53:155, 19 December 1940.

Type.— Adult female, skin and skull, MVZ 7950; from near head of Big Creek, 8000 feet altitude, Pine Forest Mountains [Humboldt County], Nevada; collected by C. H. Richardson, Jr. on 21 July 1909, original number 3198; measurements: 195-85-30.

Remarks.— Specimen in fresh summer pelage.

Eutamias amoenus monoensis Grinnell and Storer

Univ. California Publ. Zool., 17(1):3, 23 August 1916.

Type. — Adult male, skin and skull, MVZ 23380; from Warren Fork of Leevining Creek, 9200 feet altitude, Mono County, California; collected by J. Grinnell on 25 September 1915, original number 3709; measurements: 192-81-30-12, 41.1 g.

Eutamias dorsalis nidoensis Lidicker

Proc. Biol. Soc. Washington, 73:267, 30 December 1960.

Type. — Adult male, skin and skull plus stained baculum (in glycerine), MVZ 124831; from 5 miles north of Cerro Campana, 5600 feet altitude, Sierra del Nido, Chihuahua, Mexico; collected by W. Z. Lidicker, Jr. on 5 July 1959, original number 1961; measurements: 232-102-36-21, 63.9 g.

Eutamias merriami kernensis Grinnell and Storer

Univ. California Publ. Zool., 17(1):5, 23 August 1916.

Type. — Adult male, skin and skull, MVZ 15022; from Fay Creek, 4100 feet altitude, 6 miles north of Weldon, Kern County, California; collected by H. A. Carr and J. Grinnell on 13 July 1911, original number (H. A. Carr) 266; measurements: 240-110-35-17.

Remarks. — In full summer (post-breeding) pelage according to original description.

Eutamias merriami mariposae Grinnell and Storer

Univ. California Publ. Zool., 17(1):4, 23 August 1916.

Type. — Adult female, skin and skull, MVZ 21855; from El Portal, 2000 feet altitude, Mariposa County, California; collected by Walter P. Taylor on 24 November 1914, original number 7099; measurements 228-95-39-16.

Remarks. — In full winter pelage; taken in brushpile in gulch.

Eutamias minimus scrutator Hall and Hatfield

Univ. California Publ. Zool., 40(6):321, 12 February 1934.

Type. — Adult male, skin and skull, MVZ 27352; from near Blanco Mountain, 10500 feet altitude, White Mountains, Mono County, California; collected by J. Grinnell on 28 July 1917, original number 4359; measurements: 180-80-29-10, 36.7 g.

Eutamias panamintinus acrus Johnson

Univ. California Publ. Zool., 48(2):94, 24 December 1943.

Type. — Adult male, skin and skull, MVZ 86164; from 1.4 miles southeast Horse Spring, 5000 feet altitude, Kingston Range, San Bernardino County, California; collected by W. C. Russell on 9 June 1939, original number 6667; measurements: 205-90-31-17, 47.1 g.

Remarks. — In fresh summer pelage. Horse Spring = Horse Thief Spring.

Eutamias sonomae Grinnell

Univ. California Publ. Zool., 12(11):321, 20 January 1915.

Type.— Adult female, skin and skull, MVZ 20825; from 1 mile west of Guerneville, Sonoma County, California; collected by J. and H. W. Grinnell on 12 July 1913, original number (J. Grinnell) 2250; measurements: 260-112-36-17.

Remarks.— Taken in full summer pelage. Left dentary broken.

Glaucomys sabrinus flaviventris Howell

Proc. Biol. Soc. Washington, 28:112, 27 May 1915.

Type.— Adult male, skin and skull, MVZ 13319; from head of Bear Creek, 6400 feet altitude, Trinity County, California; collected by A. M. Alexander on 13 August 1911, original number 1775; measurements: 310-135-42-21.

Glaucomys sabrinus lucifugus Hall

Occ. Papers, Mus. Zool., Univ. Michigan, No. 296:1, 2 November 1934.

Type.— Adult female, skin and skull, MVZ 47613; from 12 miles east of Kamas, Summit County, Utah; collected by Virginia D. Miller on 13 July 1931, original number 101; measurements: 325-154-38-26.

Pteromys volans nebrascensis Swenk

Univ. Stud., Nebraska, 15(2):151, 25 September 1915.

Type.— Adult male, skin and skull, MVZ 31865; from Nebraska City, Otoe County, Nebraska; collected by Miss Ellen Ware on 26 November 1914, original number (Myron H. Swenk) 286; measurements: 241-115-30.5-21.5.

Remarks.— Arranged as a synonym of *Glaucomys volans volans* by Howell (N. Amer. Fauna, 44:20, 13 June 1918) but considered a valid subspecies by Hibbard (Trans. Kansas Acad. Sci., 47:74, 1944). Original tag indicates specimen was caught alive on 26 November and died two days later. The tag also differs from the original description in listing the specimen as a subadult. The skull is small, but the original description appears to give the age correctly.

Geomyidae

Geomys arenarius brevirostris Hall

Proc. Biol. Soc. Washington, 45:97, 21 June 1932.

Type.— Adult female, skin and skull, MVZ 50460; from east edge of [white] sands [9 miles west of Tularosa], Tularosa-Hot Springs Road, Otero County, New Mexico; collected by A. M. Alexander on 10 October 1931, original number 1174; measurements: 233-67-30, 165.5 g.

Geomys bursarius majusculus Swenk

Missouri Valley Fauna, Lincoln, No. 1:6, 5 December 1939.

Type. — Old adult male, skin and skull, MVZ 97913; from Lincoln, Lancaster County, Nebraska; collected by C. E. Mickel on 7 July 1919, original number 20; measurements: 352-97-40.

Geomys lutescens levisagittalis Swenk

Missouri Valley Fauna, Lincoln, No. 2:4, 1 February 1940.

Type. — Adult male, skin and skull, MVZ 97911; from Spencer [alfalfa fields in or near the Ponca River Valley], Boyd County, Nebraska; collected by C. E. Mickel on 13 August 1919, original number (Myron H. Swenk) 27; measurements: 292-91-36.

Remarks. — *Geomys bursarius levisagittalis* (Villa and Hall, Univ. Kansas Publ., Mus. Nat. Hist., 1:234, 29 November 1947).

Geomys lutescens vinaceus Swenk

Missouri Valley Fauna, Lincoln, No. 2:7, 1 February 1940.

Type. — Adult male, skin and skull, MVZ 97912; from Scottsbluff, Scottsbluff County, Nebraska; collected by M. H. Swenk on 17 August 1920, original number 32; measurements: 278-86-33.

Remarks. — A synonym of *Geomys bursarius lutescens* according to Russell and Jones (Trans. Kansas Acad. Sci., 58:513, 23 January 1956). The original tag indicates this specimen was "trapped with Maccabee trap on experiment station."

Thomomys albatus Grinnell

Univ. California Publ. Zool., 10(8):172, 7 June 1912.

Type. — Adult male, skin and skull, MVZ 10618; from [California side of the lower Colorado River at the old Hanlon Ranch] near Pilot Knob [Imperial County], California; collected by J. Dixon on 7 May 1910, original number 1396; measurements: 272-100-35.

Remarks. — *Thomomys bottae albatus* (Goldman, Proc. Biol. Soc. Washington, 48:155, 31 October 1935).

Thomomys baileyi tularosae Hall

Univ. California Publ. Zool., 38(11):411, 20 September 1932.

Type. — Adult female, skin and skull, MVZ 50444; from Cook Ranch, 1/2 mile west of Tularosa, Otero County, New Mexico; collected by L. Kellogg on 12 October 1931, original number 1195; measurements: 231-76-31, 145 g.

Remarks. — May be considered a subspecies of *Thomomys bottae* in unpublished work by J. D. Layne cited by Anderson (Syst. Zool., 15:195, 26 September 1966).

Thomomys bottae abstrusus Hall and Davis

Univ. California Publ. Zool., 40(10):391, 13 March 1935.

Type. — Adult male, skin and skull, MVZ 57613; from 2 miles southeast of Tulle Peak, 7000 feet altitude, Fish Spring Valley, Nye County, Nevada; collected by E. R. Hall on 13 July 1933, original number 3802; measurements: 245-75-31-6, 207 g.

Thomomys bottae acrirostratus Grinnell

Univ. California Publ. Zool., 40(11):408, 14 November 1935.

Type.— Mature female, skin and skull, MVZ 58940; from [valley of the] Mad River, 2700 feet altitude, [ford 7 miles] above Ruth, Trinity County, California; collected by R. M. Gilmore on 23 May 1933, original number 2663; measurements: 188-55-25-4, 76 g.

Thomomys bottae agricolaris Grinnell

Univ. California Publ. Zool., 40(11):409, 14 November 1935.

Type.— Mature female, skin and skull, MVZ 57158; from Stralock Farm, 3 miles west of Davis, Yolo County, California; received through T. I. Storer on 16 February 1933, original collector uncertain, original number (Seth B. Benson) 1912; measurements: 212-63-28-6, 120.9 g.

Remarks.— Contained six 10 mm embryos.

Thomomys bottae basilicae Benson and Tillotson

Proc. Biol. Soc. Washington, 53:93, 28 June 1940.

Type.— Adult female, skin and skull, MVZ 82221; from La Misión, 2650 feet altitude, 2 miles west of Magdalena, Sonora, Mexico; collected by Seth B. Benson and M. Delgadillo on 17 March 1938, original number (Seth B. Benson) 4835; measurements: 216-66-28-7, 110 g.

Remarks.— A new name for *Thomomys bottae occipitalis* (Benson and Tillotson, Proc. Biol. Soc. Washington, 52:151, 11 October 1939), which was preoccupied.

Thomomys bottae brevidens Hall

Univ. California Publ. Zool., 38(4):330, 27 February 1932.

Type.— Adult male, skin and skull, MVZ 49062; from Breen Creek, 7000 feet altitude, Kawich Range, Nye County, Nevada; collected by E. R. Hall on 22 September 1931, original number 3412; measurements: 232-76-31-6, 133.8 g.

Thomomys bottae cinereus Hall

Univ. California Publ. Zool., 38(4):327, 27 February 1932.

Type.— Adult female, skin and skull, MVZ 36373; from West Walker River, 4700 feet altitude, Smiths Valley, Lyon County, Nevada; collected by A. M. Alexander on 30 October 1925, original number 69; measurements: 203-61-28.

Thomomys bottae concisor Hall and Davis

Univ. California Publ. Zool., 40(10):390, 13 March 1935.

Type.— Adult female, skin and skull, MVZ 59466; from Potts Ranch, Monitor Valley, 6900 feet altitude, Nye County, Nevada; collected by W. B. Davis on 13 August 1933, original number 424; measurements: 232-77-30-5, 135.7 g.

Thomomys bottae connectens Hall

J. Washington Acad. Sci., 26(7):296, 15 July 1936.

Type. — Adult male, skin and skull, MVZ 66627; from Clawson Dairy, 4943 feet altitude, 5 miles north of Albuquerque, Bernalillo County, New Mexico; collected by A. M. Alexander on 6 May 1935, original number 2981; measurements: 256-68-35, 270.1 g.

Thomomys bottae crassus Chattin

Trans. San Diego Soc. Nat. Hist., 9(27):274, 30 April 1941.

Type. — Adult female, skin and skull, MVZ 90315; from 1-1/2 miles west of Niland, -180 feet altitude, Imperial County, California; collected by J. Chattin on 19 January 1940, original number 352; measurements: 237-81-34-7, 170 g.

Thomomys bottae curtatus Hall

Univ. California Publ. Zool., 38(4):329, 27 February 1932.

Type. — Adult male, skin and skull, MVZ 49053; from San Antonio, 5400 feet altitude, Nye County, Nevada; collected by Chester C. Lamb on 16 September 1931, original number 15792; measurements: 235-60-30-5, 155.5 g.

Thomomys bottae depressus Hall

Univ. California Publ. Zool., 38(4):326, 27 February 1932.

Type. — Adult female, skin and skull, MVZ 36970; from Dixie Meadows [at south end of Humboldt Salt Marsh], 3000 feet altitude, Churchill County, Nevada; collected by L. Kellogg and A. M. Alexander on 22 October 1926, original number (A. M. Alexander) 105; measurements: 207-60-28.

Remarks. — Original description erroneously gives altitude as 3500 feet.

Thomomys bottae detumidus Grinnell

Univ. California Publ. Zool., 40(11):405, 14 November 1935.

Type. — Mature female, skin and skull, MVZ 61336; from 1-1/2 miles south of [the town of] Pistol River, 250 feet altitude, Curry County, Oregon; collected by J. E. Hill on 11 June 1933, original number 656; measurements: 204-58-29-7.

Remarks. — Taken on "open grassy hillside."

Thomomys bottae estanciae Benson and Tillotson

Proc. Biol. Soc. Washington, 52:152, 11 October 1939.

Type. — Adult female, skin and skull, MVZ 82247; from La Estancia, 2150 feet altitude, 6 miles north of Nacori, Sonora, Mexico; collected by M. Delgadillo on 19 May 1938, original number (Seth B. Benson) 5625; measurements: 228-69-29-9-3, 155.5 g.

Remarks. — Type locality = Nacori Grande (approx. 29° 5' N lat., 110° 3' W long.), just north of Mazatan, not Nacori Chico.

Thomomys bottae fumosus Hall

Univ. California Publ. Zool., 38(4):329, 27 February 1932.

Type.— Adult male, skin and skull, MVZ 37126; from Milman Ranch [6400 feet altitude], Moores Creek, 19 miles southeast of Millett Post Office, Nye County, Nevada; collected by L. Kellogg and A. M. Alexander on 13 January 1927, original number (A. M. Alexander) 303; measurements: 236-53-32.

Thomomys bottae ingens Grinnell

Univ. California Publ. Zool., 38(10):405, 20 September 1932.

Type.— Old adult male, skin and skull, MVZ 51421; from east side levee, 298 feet altitude [2 miles due west of Millux, as shown on U. S. G. S. "Buena Vista Lake Quadrangle"], Buena Vista Lake, Kern County, California; collected by J. Grinnell on 26 February 1932, original number 7065; measurements 275-85-35-6, 278 g.

Remarks.— Original description gives altitude as 290 feet. In full breeding condition when collected according to original description.

Thomomys bottae lacrymalis Hall

Univ. California Publ. Zool., 38(4):328, 27 February 1932.

Type.— Adult female, skin and skull, MVZ 38427; from Arlemont [=Chiatovich Ranch, Fish Lake Valley], 4900 feet altitude, Esmeralda County, Nevada; collected by A. M. Alexander and L. Kellogg on 22 May 1927, original number (A. M. Alexander) 390; measurements: 222-54-28.

Thomomys bottae latus Hall and Davis

Univ. California Publ. Zool., 40(10):393, 13 March 1935.

Type.— Adult female, skin and skull, MVZ 46074; from Cherry Creek, 6500 feet altitude, White Pine County, Nevada; collected by Chester C. Lamb on 12 September 1930, original number 12936; measurements: 220-69-29-5, 161.3 g.

Thomomys bottae lucidus Hall

Proc. Biol. Soc. Washington, 45:67, 2 April 1932.

Type.— Adult female, skin and skull, MVZ 39119; from Las Palmas Canyon, 200 feet altitude, west side of Laguna Salada, 15 miles south of north end [north of 32° N latitude], Baja California, Mexico; collected by J. Elton Green on 30 October 1927, original number 881; measurements: 202-64-30-5, 73 g.

Thomomys bottae lucrificus Hall and Durham

Proc. Biol. Soc. Washington, 51:15, 18 February 1938.

Type.— Adult male, skin and skull, MVZ 70602; from Eastgate, Churchill County, Nevada; collected by W. C. Russell on 15 May 1936, original number 4392; measurements: 260-88-33-6, 198.5 g.

Thomomys bottae nanus Hall

Univ. California Publ. Zool., 38(4):331, 27 February 1932.

Type. — Adult female, skin and skull, MVZ 49077; from 5-1/2 miles northwest Whiterock Spring, 7200 feet altitude, south end of Belted Range, Nye County, Nevada; collected by Chester C. Lamb on 27 September 1931, original number 15921; measurements: 193-57-29-5, 80.6 g.

Thomomys bottae nasutus Hall

Proc. Biol. Soc. Washington, 45:96, 21 June 1932.

Type. — Adult female, skin and skull, MVZ 50343; from west fork Black River, 7750 feet altitude, Apache County, Arizona; collected by A. M. Alexander on 14 June 1931, original number 892; measurements: 205-52-28.

Remarks. — A synonym of *Thomomys bottae fulvus* according to Goldman (Proc. Biol. Soc. Washington, 48:156, 31 October 1935). Original description erroneously gives altitude as 7550 feet.

Thomomys bottae piutensis Grinnell and Hill

Proc. Biol. Soc. Washington, 49:103, 22 August 1936.

Type. — Adult female, skin and skull, MVZ 60051; from French Gulch, 6700 feet altitude [2-1/2 miles northwest of Claraville], Piute Mountains, Kern County, California; collected by David S. MacKaye on 23 October 1933, original number 259; measurements: 210-65-28-7, 100 g.

Thomomys bottae ruidosae Hall

Proc. Biol. Soc. Washington, 45:96, 21 June 1932.

Type. — Adult female, skin and skull, MVZ 50431; from Ruidoso, 6700 feet altitude, Lincoln County, New Mexico; collected by L. Kellogg on 30 September 1931, original number 1158; measurements: 200-60-29, 98 g.

Thomomys bottae rupestris Chattin

Trans. San Diego Soc. Nat. Hist., 9(27):272, 30 April 1941.

Type. — Adult female, skin and skull, MVZ 84683; from 2 miles east of Clemens Well, 1131 feet altitude, Riverside County, California; collected by A. M. Alexander on 1 January 1939, original number 3295; measurements: 198-64-30, 80.2 g.

Thomomys bottae saxatilis Grinnell

Proc. Biol. Soc. Washington, 47:193, 2 October 1934.

Type. — Mature female, skin and skull, MVZ 63354; from 1 mile north of Susanville, 4400 feet altitude, Lassen County, California; collected by J. and H. W. Grinnell on 31 May 1934, original number (J. Grinnell) 7341; measurements: 200-57-27-3.5, 103.6 g.

Thomomys bottae siccovallis Huey

Trans. San Diego Soc. Nat. Hist., 10(14):258, 31 August 1945.

Type.— Adult female, skin and skull, MVZ 37673; from El Cajón Canyon, 3200 feet altitude, east base of San Pedro Mártir Mountains [30° 54' N latitude, 115° 10' W longitude], Baja California, Mexico; collected by Raymond M. Gilmore on 3 June 1926, original number 369; measurements: 209-59-28-4, 108.5 g.

Remarks.— Huey (1945) notes the type locality is "a very secluded canyon on the desert side of the Sierra San Pedro Mártir."

Thomomys bottae silvifugus Grinnell

Univ. California Publ. Zool., 40(11):406, 14 November 1935.

Type.— Mature female, skin and skull, MVZ 58958; from [near] Coyote Peak, 3000 feet altitude, 16 miles due east of Patrick's Point, Humboldt County, California; collected by R. M. Gilmore and W. H. Behle on 14 June 1933, original number (R. M. Gilmore) 2808; measurements: 205-65-26-3, 118.5 g.

Thomomys bottae trumbullensis Hall and Davis

Proc. Biol. Soc. Washington, 47:51, 9 February 1934.

Type.— Adult male, skin and skull, MVZ 58588; from 3 miles south of Nixon Spring, Mount Trumbull, Mohave County, Arizona; collected by Seth B. Benson on 26 May 1933, original number 2005; measurements: 237-75-29-7, 177 g.

Thomomys bottae vescus Hall and Davis

Univ. California Publ. Zool., 40(10):389, 13 March 1935.

Type.— Adult female, skin and skull, MVZ 57606; from south slope of Mount Jefferson, 9000 feet altitude, Toquima Range, Nye County, Nevada; collected by E. R. Hall and S. D. Durrant on 3 July 1933, original number (E. R. Hall) 3729; measurements: 205-66-28-5, 96 g.

Thomomys chrysonotus Grinnell

Univ. California Publ. Zool., 10(8):174, 7 June 1912.

Type.— Young adult male, skin and skull, MVZ 10617; from Ehrenberg, Colorado River [Yuma County], Arizona; collected by Frank Stephens on 27 March 1910, original number 2540; measurements: 217-73-30-5.

Remarks.— *Thomomys bottae chrysonotus* (Goldman, Proc. Biol. Soc. Washington, 48:156, 31 October 1935).

Thomomys diaboli Grinnell

Univ. California Publ. Zool., 12(9):313, 21 November 1914.

Type.— Adult female, skin and skull, MVZ 14165; from Sweeney's Ranch [near Sweeney Hill], 22 miles [by road] southwest of Los Banos, in hills of Diablo Range, Merced County, California; collected by H. A. Carr and C. H. Richardson on 2 April 1911, original number (H. A. Carr) 108; measurements: 180-60-25.

Remarks.— *Thomomys bottae diaboli* (Bailey, N. Amer Fauna, 39:51, 15 November 1915).

Thomomys falcifer Grinnell
Univ. California Publ. Zool., 30(6):180, 10 December 1926.

Type.— Adult male, skin and skull, MVZ 35043; from Bell's Ranch, Reese River Valley, 6890 feet altitude, Nye County, Nevada; collected by A. M. Alexander on 3 June 1925, original number 15; measurements: 218-56-27.

Remarks.— *Thomomys talpoides falcifer* (Goldman, J. Mamm., 20:234, 15 May 1939). Miss Alexander kept several different field catalogs prior to initiating a continuous series of field numbers in 1926, thus explaining the low original number of this type when compared to those from Alaska in the years 1907 and 1909.

Thomomys harquahalae Grinnell and Hill
J. Mamm., 17(1):7, 14 February 1936.

Type.— Adult female, skin and skull, MVZ 62085; from Ranegras Plain, 10 miles west of Hope, Yuma County, Arizona; collected by L. Kellogg on 27 February 1934, original number 2856; measurements: 235-69-33-5, 168 g.

Remarks.— *Thomomys umbrinus harquahalae* (Hall and Kelson, The mammals of North America, p. 425. The Ronald Press, Inc., 31 March 1959), which should now be considered a subspecies of *Thomomys bottae*, following the work of Patton and Dingman (J. Mamm., 49:1, 20 February 1968). The altitude of the type locality is given variously as "approximately 950 feet" (in the original description) and "about 1250 feet" (in Miller and Kellogg, 1955). Hooper (Misc. Publ. Mus. Zool., Univ. Michigan, No. 52:12, 1941) further defines the type locality as "10 miles west-southwest of Vicksberg."

Thomomys infrapallidus Grinnell
Univ. California Publ. Zool., 12(9):314, 21 November 1914.

Type.— Old adult male, skin and skull, MVZ 14181; from [Pimental Ranch, a Miller and Lux Company ranch house] 7 miles southeast of Simmler, Carrizo Plains, San Luis Obispo County, California; collected by H. S. Swarth on 25 May 1911, original number 9138; measurements: 248-76-34.

Remarks.— *Thomomys bottae infrapallidus* (Bailey, N. Amer. Fauna, 39:55, 15 November 1915).

Thomomys jacinteus Grinnell and Swarth
Proc. California Acad. Sci., 4th Ser., 4:154, 30 December 1914.

Type.— Adult male, skin and skull, MVZ 21235; from Round Valley, 9000 feet altitude, San Jacinto Mountains, Riverside County, California; collected by H. S. Swarth on 15 September 1914, original number 10012; measurements: 240-82-32-5.

Remarks.— *Thomomys bottae jacinteus* (Goldman, Proc. Biol. Soc. Washington, 48:155, 31 October 1935).

Thomomys melanotis Grinnell

Univ. California Publ. Zool., 17(14):425, 25 April 1918.

Type.— Adult male, skin and skull, MVZ 26499; from Big Prospector Meadow, White Mountains, 10500 feet altitude, Mono County, California; collected by A. C. Shelton on 27 July 1917, original number 3402; measurements: 225-72-30-4, 147.9 g.

Remarks.— *Thomomys bottae melanotis* (Hall, Univ. California Publ. Zool., 38:328, 27 February 1932).

Thomomys monticola premaxillaris Grinnell

Univ. California Publ. Zool., 12(9):312, 21 November 1914.

Type.— Old adult male, skin with complete skeleton, MVZ 20242; from 2 miles south of South Yolla Bolly Mountain [altitude about 7500 feet], Tehama County, California; collected by G. F. Ferris on 6 August 1913, original number 166; measurements: 215-59-27.

Remarks.— A synonym of *Thomomys monticola monticola* according to Goldman (J. Washington Acad. Sci., 33:146, 15 May 1943).

Thomomys nigricans puertae Grinnell

Univ. California Publ. Zool., 12(9):315, 21 November 1914.

Type.— Young adult male, skin and skull, MVZ 7511; from La Puerta [Mason's Ranch, 5 miles west of Vallecito, on the old overland (Butterfield) stage road, lower end of the La Puerta Valley], San Diego County, California; collected by Frank Stephens on 31 May 1909, original number 1974; measurements: 203-80-28-6.

Remarks.— *Thomomys bottae puertae* (Bailey, N. Amer. Fauna, 39:58, 15 November 1915).

Thomomys perpallidus albicaudatus Hall

Univ. California Publ. Zool., 32(6):444, 8 July 1930.

Type.— Adult male, skin and skull, MVZ 43971; from Provo, 4510 feet altitude, Utah County, Utah; collected by A. M. Alexander on 17 October 1929, original number 506; measurements: 237-60-32.

Remarks.— *Thomomys bottae albicaudatus* (Goldman, Proc. Biol. Soc. Washington, 48:156, 31 October 1935).

Thomomys perpallidus amargosae Grinnell

Univ. California Publ. Zool., 21(6):239, 7 November 1921.

Type.— Old adult male, skin and skull, MVZ 26485; from Shoshone, 1560 feet altitude, [on the Amargosa "River"], T 22 N, R 7 E SBM, Inyo County, California; collected by Tracy I. Storer on 14 May 1917, original number 1777; measurements: 259-81-32.

Remarks.— *Thomomys bottae amargosae* (Hall, Univ. California Publ. Zool., 38:328, 27 February 1932).

Thomomys perpallidus aureiventris Hall

Univ. California Publ. Zool., 32(6):444, 8 July 1930.

Type. — Adult male, skin and skull, MVZ 43980; from [Fehlman Ranch, 3 miles north of] Kelton, 4225 feet altitude, Box Elder County, Utah; collected by L. Kellogg on 27 September 1929, original number 451; measurements: 240-72-32.

Remarks. — *Thomomys bottae aureiventris* (Goldman, Proc. Biol. Soc. Washington, 48:156, 31 October 1935).

Thomomys perpallidus centralis Hall

Univ. California Publ. Zool., 32(6):445, 8 July 1930.

Type. — Adult male, skin and skull, MVZ 41688; from 2-1/2 east of Baker [1-1/4 miles west of Nevada-Utah boundary on 39th parallel], 5700 feet altitude, White Pine County, Nevada; collected by E. R. Hall on 30 May 1929, original number 2683; measurements: 242-75-31.5-5.7, 197.5 g.

Remarks. — *Thomomys bottae centralis* (Hall, Univ. California Publ. Zool., 38:333, 27 February 1932).

Thomomys perpallidus depauperatus Grinnell and Hill

J. Mamm., 17(1):4, 14 February 1936.

Type. — Adult female, skin and skull, MVZ 62079; from [east base of] Tinajas Altas Mountains [7 miles south of Raven Butte], 1150 feet altitude, Yuma County, Arizona; collected by A. M. Alexander on 17 January 1934, original number 2704; measurements: 188-60-28-4, 66.1 g.

Remarks. — *Thomomys bottae depauperatus* (Huey, Trans. San Diego Soc. Nat. Hist., 8:353, 15 June 1937). But note that Goldman (Proc. Biol. Soc. Washington, 48:153, 31 October 1935) had earlier stated that all named subspecies of *Thomomys perpallidus* should be considered subspecies of *Thomomys bottae*.

Thomomys perpallidus mohavensis Grinnell

Univ. California Publ. Zool., 17(14):427, 25 April 1918.

Type. — Adult male, skin and skull, MVZ 4639; from [Mohave River bottom, 2700 feet altitude, near] Victorville [San Bernardino County], California; collected by J. Grinnell and J. Dixon on 26 December 1904, original number (J. Grinnell) 906; measurements: 235-75-31.

Remarks. — *Thomomys bottae mohavensis* (Goldman, Proc. Biol. Soc. Washington, 48:155, 31 October 1935).

Thomomys perpallidus riparius Grinnell and Hill

J. Mamm., 17(1):4, 14 February 1936.

Type. — Adult female, skin and skull, MVZ 61985; from Blythe, Riverside County, California; collected by A. M. Alexander on 20 February 1934, original number 2831; measurements: 228-76-31-5, 135 g.

Remarks. — *Thomomys bottae riparius* (Chattin, Trans. San Diego Soc. Nat. Hist., 9(27):268, 30 April 1941).

Thomomys providentialis Grinnell

Univ. California Publ. Zool., 38(1):1, 17 October 1931.

Type.— Adult female, skin and skull, MVZ 31450; from Purdy, 4500 feet altitude, 6 miles southeast of New York Mountain [Providence Range], San Bernardino County, California; collected by J. Grinnell on 26 April 1920, original number 5249; measurements: 205-65-27-3, 114.3 g.

Remarks.— *Thomomys bottae providentialis* (Hall and Davis, Univ. California Publ. Zool., 40(10):400, 13 March 1935).

Thomomys quadratus gracilis Durrant

Bull. Univ. Utah, 29(6):3, [Biol. Ser., 30(10):3], 28 February 1939.

Type.— Adult male, skin and skull, MVZ 44866; from Pine Canyon, 6600 feet altitude, 17 miles northwest of Kelton, Raft River Mountains, Box Elder County, Utah; collected by A. M. Alexander on 12 July 1930, original number 676; measurements: 210-54-28.

Remarks.— *Thomomys talpoides gracilis* (Durrant, Bull. Univ. Utah, 30(5):6, [Biol. Ser., 5(4):6], 24 October 1939).

Thomomys quadratus wallowa Hall and Orr

Proc. Biol. Soc. Washington, 46:41, 24 March 1933.

Type.— Adult male, skin and skull, MVZ 54701; from Catherine Creek, 3500 feet altitude, 7 miles east of Telocaset, Union County, Oregon; collected by Robert T. Orr on 29 June 1932, original number 570; measurements: 188-54-25-6, 75.5 g.

Remarks.— *Thomomys talpoides wallowa* (Goldman, J. Mamm., 20:234, 15 May 1939). Arranged as a synonym of *Thomomys talpoides fuscus* by Bailey (N. Amer. Fauna, 55:259, 29 August 1936), but considered valid by Dalquest and Scheffer (Amer. Nat., 78:437, September 1944).

Thomomys relictus Grinnell

Univ. California Publ. Zool., 30(1):2, 18 August 1926.

Type.— Old adult male, skin and skull, MVZ 35271; from [the valley of the Susan River] 2 miles south of Susanville, Lassen County, California; collected by J. Dixon on 5 June 1925, original number 8581; measurements: 280-80-38-6, 325.1 g.

Remarks.— *Thomomys townsendii relictus* (Grinnell, Univ. California Publ. Zool., 40:137, 26 September 1933).

Thomomys solitarius Grinnell

Univ. California Publ. Zool., 30(6):177, 10 December 1926.

Type.— Subadult male, skin and skull, MVZ 36378; from Finger-rock Wash, 5400 feet altitude, Stewart Valley, Mineral County, Nevada; collected by A. M. Alexander and L. Kellogg on 9 October 1925, original number 54; measurements: 205-66-28.

Remarks.— *Thomomys bottae solitarius* (Hall, Univ. California Publ. Zool., 38(4):333, 27 February 1932).

Thomomys talpoides aequalidens Dalquest

Murrelet, 23(1):3, 14 May 1942.

Type. — Adult male, skin and skull, MVZ 66403; from Abel Place, 2200 feet altitude, 6 miles south-southeast of Dayton [Columbia County], Washington; collected by S. H. Lyman on 6 April 1934, original number M-28; measurements: 206-58-27.

Thomomys talpoides cheyennensis Swenk

Missouri Valley Fauna, Lincoln, No. 4:5, 1 March 1941.

Type. — Old adult male, skin and skull, MVZ 97909; from 2 miles south of Dalton, Cheyenne County, Nebraska; collected by C. E. Mickel on 26 June 1919, original number 18; measurements: 225-66-29.

Thomomys talpoides confinis Davis

Murrelet, 18(1-2):25, 4 September 1937.

Type. — Adult female, skin and skull, MVZ 77301; from Gird Creek [near Hamilton], Rivalli County, Montana; collected by William L. Jellison and Nick Kramis on 10 October 1934, original number (W. L. Jellison) 10189; measurements: 18-50-23 mm (sic).

Thomomys talpoides devexus Hall and Dalquest

Murrelet, 20(1):3, 30 April 1939.

Type. — Adult male, skin and skull, MVZ 84517; from 1 mile west-southwest of Neppel, Grant County, Washington; collected by W. W. Dalquest on 30 May 1938, original number 688; measurements: 179-49-26-7.

Thomomys talpoides immunis Hall and Dalquest

Murrelet, 20(1):4, 30 April 1939.

Type. — Subadult male, skin and skull, MVZ 84519; from 5 miles south of Trout Lake, Klickitat County, Washington; collected by W. W. Dalquest on 27 July 1937, original number 395; measurements: 192-60-27-5.

Thomomys talpoides pierreicolus Swenk

Missouri Valley Fauna, Lincoln, No. 4:2, 1 March 1941.

Type. — Old adult male, skin and skull, MVZ 97910; from Wayside, Dawes County, Nebraska; collected by C. E. Mickel on 24 September 1919, original number 38; measurements: 217-75-26.

Thomomys talpoides yakimensis Hall and Dalquest

Murrelet, 20(1):4, 30 April 1939.

Type. — Adult female, skin and skull, MVZ 84518; from Selah, Yakima County, Washington; collected by Peggy Burgner on 27 November 1938, original number (W. W. Dalquest) 1077; measurements: 186-55-25-6.

Thomomys townsendii bachmani Davis

J. Mamm., 18(2):150, 14 May 1937.

Type.— Adult female, skin and skull, MVZ 7855; from Quinn River Crossing, 4100 feet altitude, Humboldt County, Nevada; collected by C. H. Richardson, Jr. and W. P. Taylor on 29 May 1909, original number (C. H. Richardson) 2704; measurements: 262-81-45? (sic).

Remarks.— Original label lost, current label indicates date of collection was 27 May 1909. However the collector's field catalog confirms 29 May 1909 as the correct date.

Thomomys townsendii elkoensis Davis

J. Mamm., 18(2):151, 14 May 1937.

Type.— Adult female, skin and skull, MVZ 70583; from Evans, Eureka County, Nevada; collected by W. C. Russell on 30 May 1936, original number 4614; measurements: 255-75-35-6, 228.2 g.

Thomomys townsendii owyhensis Davis

J. Mamm., 18 (2):154, 14 May 1937.

Type.— Adult female, skin and skull, MVZ 67490; from Castle Creek, 8 miles south of Oreana, Owyhee County, Idaho; collected by W. B. Davis on 22 May 1935, original number 1260; measurements: 255-77-33-8.

Remarks.— Original label indicates specimen was taken in "clay-loam soil in alfalfa field."

Thomomys townsendii similis Davis

J. Mamm., 18(2):155, 14 May 1937.

Type.— Adult female, skin and skull, MVZ 46507; from West Pocatello River bottom, Pocatello, Bannock County, Idaho; collected by W. B. Whitlow on 10 November 1928, original number 182; measurements: 260-83-35-6.

Thomomys umbrinus pullus Hall and Villa

Univ. Kansas Publ., Mus. Nat. Hist., 1(14):251, 26 July 1948.

Type.— Adult male, skin and skull, MVZ 100151; from 5 miles south of Pátzcuaro, 7800 feet altitude, Michoacán, Mexico; collected by H. H. Hall on 10 March 1943, original number 117; measurements: 182-53-27-6, 87.8 g.

Heteromyidae

Perognathus amplus ammodytes Benson

Proc. Biol. Soc. Washington, 46:110, 27 April 1933.

Type.— Adult male, skin and skull, MVZ 55774; from [2 miles south of] Cameron, 4200 feet altitude, Coconino County, Arizona; collected by L. Kellogg on 8 August 1932, original number 1620; measurements: 127-66-20-5, 9.6 g.

Perognathus amplus cineris Benson

Proc. Biol. Soc. Washington, 46:109, 27 April 1933.

Type. — Adult male, skin and skull, MVZ 55771; from [near the] Wupatki Ruins, Wupatki National Monument [about 27 miles northeast of Flagstaff], Coconino County, Arizona; collected by A. M. Alexander on 12 October 1932, original number 1932; measurements: 143-78-20-4, 10.3 g.

Remarks. — Original description indicated this form is "known only from ground covered with black sand (volcanic cinders) in the vicinity of the Wupakti Ruins, but possibly also occurring in similar places elsewhere in the San Francisco Mountain volcanic field."

Perognathus californicus bensoni von Bloeker

Proc. Biol. Soc. Washington, 51:197, 23 December 1938.

Type. — Subadult male, skin and skull, MVZ 81579; from Stonewall Creek, 1300 feet altitude, 6-3/10 miles northeast of Soledad, Monterey County, California; collected by Jack C. von Bloeker, Jr. on 16 June 1937, original number 7771; measurements: 197-107-25-12.

Perognathus californicus bernardinus Benson

Univ. California Publ. Zool., 32(7):449, 6 September 1930.

Type. — Adult male, skin and skull, MVZ 44094; from 2 miles east of Strawberry Peak [5750 feet altitude], San Bernardino Mountains [San Bernardino County], California; collected by Laurence M. Huey on 19 September 1920, original number H631; measurements: 227-131-28-11.

Remarks. — Originally collected for Donald R. Dickey.

Perognathus californicus marinensis von Bloeker

Proc. Biol. Soc. Washington, 51:199, 23 December 1938.

Type. — Adult male, skin and skull, MVZ 81550; from Indian Harbor, 50 feet altitude, 1-1/2 miles south of Marina, Monterey County, California; collected by Jack C. von Bloeker, Jr. on 27 May 1937, original number 7496; measurements: 203-114-26-14.

Perognathus flavescens olivaceogriseus Swenk

Missouri Valley Fauna, Lincoln. No. 3:6, 5 June 1940.

Type. — Adult male, skin and skull, MVZ 97914; from 5 miles east of Chadron, Dawes County, Nebraska; collected by Leroy M. Gates on 26 October 1915, original number 2; measurements: 124-53-16.5-4 (from crown).

Remarks. — *Perognathus fasciatus olivaceogriseus* (Jones, Univ. Kansas Publ., Mus. Nat. Hist., 5:520, 1 August 1953). Original description further defines type locality as "farm along Little Bordeaux Creek, 3 miles east of Chadron, section 14, T 33 N, R 48 W, composed largely of Laurel and Tripp very fine sandy loam."

Perognathus formosus cinerascens Nelson and Goldman

Proc. Biol. Soc. Washington, 42:105, 25 March 1929.

Type. — Female in adult pelage but molars unworn, skin and skull, MVZ 37685; from San Felipe, northeastern Baja California, Mexico; collected by Chester C. Lamb on 10 April 1926, original number 5772; measurements: 155-75-22-8, 15.8 g.

Remarks. — Skull broken in half behind frontals.

Perognathus formosus incolatus Hall

Proc. Biol. Soc. Washington, 54:56, 20 May 1941.

Type. — Adult female, skin and skull, MVZ 78855; from 2 miles west of Smith Creek Cave, 6300 feet altitude, Mount Moriah, White Pine County, Nevada; collected by Lee W. Arnold on 18 June 1937, original number 360; measurements: 190-103-23-12, 21.5 g.

Perognathus formosus melanurus Hall

Proc. Biol. Soc. Washington, 54:57, 20 May 1941.

Type. — Adult male, skin and skull, MVZ 73442; from 40° 28' N latitude, 6 miles east of California line, 4000 feet altitude, Washoe County, Nevada; collected by E. R. Hall on 7 July 1936, original number 5070; measurements: 199-108-25.3-11, 21.2 g.

Remarks. — Original description erroneously gives sex as female.

Perognathus inornatus sillimani von Bloeker

Proc. Biol. Soc. Washington, 50:154, 10 September 1937.

Type. — Adult female, skin and skull, MVZ 74682; from west side of Arroyo Seco Wash, 4 miles south of Soledad, 150 feet altitude, Monterey County, California; collected by Jack C. von Bloeker, Jr. on 3 August 1936, original number 6997; measurements: 145-73-19-8.

Perognathus intermedius crinitus Benson

Proc. Biol. Soc. Washington, 47:199, 2 October 1934.

Type. — Adult male, skin and skull, MVZ 55883; from 2.6 miles west of the Wupatki Ruins National Monument, Coconino County, Arizona; collected by L. Kellogg on 8 October 1932, original number 1889; measurements: 177-101-23.5-5, 13.5 g.

Perognathus intermedius rupestris Benson

Univ. California Publ. Zool., 38(5):337, 14 April 1932.

Type. — Young adult male, skin and skull, MVZ 50595; from [that part of the] lava beds nearest to Kenzin, Dona Ana County, New Mexico; collected by A. M. Alexander on 24 October 1931, original number 1225; measurements: 169-92-20.5-4, 12.9 g.

Perognathus intermedius umbrosus Benson

Proc. Biol. Soc. Washington, 47:200, 2 October 1934.

Type.— Adult male, skin and skull, MVZ 55964; from Camp Verde, Yavapai County, Arizona; collected by L. Kellogg on 3 October 1932, original number 1863; measurements: 173-99-23-5, 16.7 g.

Perognathus longimembris arcus Benson

Univ. California Publ. Zool., 40(14):451, 31 December 1935.

Type.— Adult female, skin and skull, MVZ 58624; from Rainbow Bridge, 4000 feet altitude, San Juan County, Utah; collected by Seth B. Benson on 23 June 1933, original number 2178; measurements: 145-86-19-8, 8.5 g.

Perognathus longimembris cantwelli von Bloeker

Proc. Biol. Soc. Washington, 45:128, 9 September 1932.

Type.— Adult female, skin and skull, MVZ 74680; from Hyperion, 50 feet altitude [El Segundo], Los Angeles County, California; collected by Jack C. von Bloeker, Jr. on 5 September 1931, original number 1378; measurements: 133-72-18-7.

Remarks.— A synonym of *Perognathus longimembris pacificus* according to Huey (Trans. San Diego Soc. Nat. Hist., 9(11):49, 31 August 1939). Type described while in collection of Jack C. von Bloeker, Jr.

Perognathus longimembris gulosus Hall

Proc. Biol. Soc. Washington, 54:55, 20 May 1941.

Type.— Adult female, skin and skull, MVZ 78764; from [1/4 mile south of] Smith Creek Cave, 5800 feet altitude, Mount Moriah, White Pine County, Nevada; collected by Lee W. Arnold on 4 June 1937, original number 259; measurements: 132-72-17-7, 8.9 g.

Perognathus longimembris neglectus Taylor

Univ. California Publ. Zool., 10(6):155, 21 May 1912.

Type.— Adult male, skin and skull, MVZ 14526; from McKittrick, 1111 feet altitude, Kern County, California; collected by H. S. Swarth and W. L. Chandler on 18 May 1911, original number (H. S. Swarth) 8986; measurements: 157-77-22-6.

Remarks.— *Perognathus inornatus neglectus* (Osgood, Proc. Biol. Soc. Washington, 31:96, 29 June 1918).

Perognathus longimembris psammophilus von Bloeker

Proc. Biol. Soc. Washington, 50:153, 10 September 1937.

Type.— Subadult female, skin and skull, MVZ 74681; from west side of Arroyo Seco Wash, 150 feet altitude, 4 miles south of Soledad, Monterey County, California; collected by Jack C. von Bloeker, Jr. on 5 June 1936, original number 6209; measurements: 130-62-18-7.

Perognathus longimembris tularensis Richardson

J. Mamm., 18(4):510, 14 November 1937.

Type. — Adult male, skin and skull, MVZ 74668; from 1 mile west of Kennedy Meadows, 6000 feet altitude, South Fork of the Kern River, Tulare County, California; collected by William B. Richardson on 14 August 1936, original number 2004; measurements: 127-68-19-7.

Perognathus parvus trumbullensis Benson

Proc. Biol. Soc. Washington, 50:181, 28 October 1937.

Type. — Young adult male, skin and skull, MVZ 60929; from Nixon Spring, 6250 feet altitude, Mount Trumbull, Mohave County, Arizona; collected by by A. M. Alexander on 26 May 1933, original number 2182; measurements: 174-90-23-9, 17.2 g.

Perognathus spinatus lambi Benson

Univ. California Publ. Zool., 32(7):452, 6 September 1930.

Type. — Young adult female, skin and skull, MVZ 42938; from [San Gabriel] Espíritu Santo Island, Baja California, Mexico; collected by Chester C. Lamb on 19 January 1929, original number 10264; measurements: 175-105-23-6.

Perognathus xanthonotus Grinnell

Proc. Biol. Soc. Washington, 25:128, 31 July 1912.

Type. — Adult male, skin and skull, MVZ 16154; from Freeman Canyon [4900 feet altitude], east slope of Walker Pass, Kern County, California; collected by H. A. Carr on 27 June 1911, original number 111; measurements: 170-85-23.5.

Remarks. — Occipital region of skull damaged.

Microdipodops megacephalus ambiguus Hall

Field Mus. Nat. Hist., Publ. 511, Zool. Ser., 27:252, 8 December 1941.

Type. — Adult male, skin and skull, MVZ 73840; from 1-1/4 miles north of Sulphur, 4050 feet altitude, Humboldt County, Nevada; collected by E. R. Hall on 25 July 1936, original number 5285; measurements: 162-89-25-9.8, 13.2 g.

Microdipodops megacephalus medius Hall

Field Mus. Nat. Hist., Publ. 511, Zool. Ser., 27:256, 8 December 1941.

Type. — Adult female, skin and skull, MVZ 73890; from 3 miles south of Vernon, 4250 feet altitude, Pershing County, Nevada; collected by E. R. Hall on 28 July 1936, original number 5324; measurements: 165-89-25-10.5.

Microdipodops megacephalus nasutus Hall

Field Mus. Nat., Hist., Publ. 511, Zool. Ser., 27:251, 8 December 1941.

Type. — Adult female, skin and skull, MVZ 40439; from Fletcher, 6098 feet altitude, Mineral County, Nevada; collected by L. Kellogg on 22 July 1928, original number 374; measurements: 158-88-25.

Microdipodops megacephalus nexus Hall

Field Mus. Nat. Hist., Publ. 511, Zool. Ser., 27:257, 8 December 1941.

Type.— Adult male, skin and skull, MVZ 70917; from 3 miles south of Izenhood, Lander County, Nevada; collected by W. C. Russell on 22 May 1936, original number 4466; measurements: 167-99-25-10.

Microdipodops megacephalus paululus Hall and Durrant

Murrelet, 22(1):5, 30 April 1941.

Type.— Subadult male, skin and skull, MVZ 74660; from Pine Valley, 5000 feet altitude, 50 miles west of Milford, Millard County, Utah; collected by S. D. Durrant on 17 July 1936, original number 955; measurements: 147-76-24-11.

Remarks.— The original description further defines the type locality as "1/2 mile east of headquarters building of the Desert Range Experiment Station of the United States Forest Service, Sec. 33, T 25 S, R 17 W, Salt Lake B. M."

Microdipodops megacephalus sabulonis Hall

Proc. Biol. Soc. Washington, 54:59, 20 May 1941.

Type.— Adult male, skin and skull, MVZ 49381; from 5 miles southeast of the Kawich Post Office [Kawich Valley]. 5400 feet altitude, Nye County, Nevada; collected by R. T. Orr on 27 September 1931, original number 384; measurements: 155-83-25-10, 11.3 g.

Microdipodops pallidus albiventer Hall and Durrant

J. Mamm., 18(3):357, 14 August 1937.

Type.— Adult male, skin and skull, MVZ 52803; from Desert Valley, 5300 feet altitude, 21 miles west of Panaca, Lincoln County, Nevada; collected by W. C. Russell on 30 May 1932, original number 2188; measurements: 150-80-24-9.5, 11.6 g.

Remarks.— *Microdipodops megacephalus albiventer* (Hall, Field Mus. Nat. Hist., Publ. 511, Zool. Ser., 27:263, 8 December 1941).

Microdipodops pallidus ammophilus Hall

Field Mus. Nat. Hist., Publ. 511, Zool. Ser., 27:273, 8 December 1941.

Type.— Adult female, skin and skull, MVZ 58208; from Able Spring, 12-1/2 miles south Locks Ranch, 5000 feet altitude, Railroad Valley, Nye County, Nevada; collected by E. R. Hall on 29 July 1933, original number 3934; measurements: 162-90-26.2-11, 14.6 g.

Microdipodops pallidus purus Hall

Field Mus. Nat. Hist., Publ. 511, Zool. Ser., 27:273, 8 December 1941.

Type.— Adult male, skin and skull, MVZ 52753; from 14-1/2 miles south of Groom Baldy, Lincoln County, Nevada; collected by E. R. Hall and H. R. Poultney on 1 June 1932, original number (E. R. Hall) 3618; measurements: 160-88-26-10, 12.9 g.

Microdipodos (sic) pallidus ruficollaris Hall

Proc. Biol. Soc. Washington, 54:60, 20 May 1941.

Type. — Adult female, skin and skull, MVZ 49254; from 5 miles southeast of the Kawich Post Office [Kawich Valley], 5400 feet altitude, Nye County, Nevada; collected by R. T. Orr on 25 September 1931, original number 367; measurements: 160-90-25-9, 12.0 g.

Microdipodops polionotus Grinnell

Univ. California Publ. Zool., 12(7):302, 15 April 1914.

Type. — Adult male, skin with complete skeleton, MVZ 17031; from McKeever's Ranch, 2 miles south of Benton Station, 5200 feet altitude, Mono County, California; collected by C. D. Hollinger on 10 July 1912, original number 184; measurements: 145-80-24-9.

Remarks. — *Microdipodops megacephalus polionotus* (Hall, Field Mus. Nat. Hist., Publ. 511, Zool. Ser., 27:251, 8 December 1941). Occipital region of skull broken. The original description further defines the type locality as "a sandy, sagebrush flat on McKeever Ranch, two miles south of the railroad station at Benton."

Dipodomys agilis fuscus Boulware

Univ. California Publ. Zool., 46(7):393, 16 September 1943.

Type. — Adult male, skin and skull plus baculum, MVZ 84845; from 1-1/2 miles due north of La Purisima Mission, 600 feet altitude, Santa Barbara County, California; collected by R. M. Bond on 16 February 1939, original number 132; measurements: 305-181-43-16.

Remarks. — The original description erroneously gives the original number as 162 and the type locality as 2-1/2 miles north of La Purisima Mission.

Dipodomys berkeleyensis Grinnell

Proc. Biol. Soc. Washington, 32:204, 31 December 1919.

Type. — Adult male, skin with complete skeleton, MVZ 28729; from [top of hill at head of Dwight Way], Berkeley [Alameda County], California; collected by J. Grinnell and D. D. McLean on 6 October 1918, original number (J. Grinnell) 4815; measurements: 301-180-41-12, 77 g.

Remarks. — *Dipodomys heermanni berkeleyensis* (Grinnell, J. Mamm., 2:95, 2 May 1921).

Dipodomys californicus eximius Grinnell

Proc. Biol. Soc. Washington, 32:205, 31 December 1919.

Type. — Adult male, skin and skull, MVZ 18347; from Marysville Buttes, 300 feet altitude, 3 miles northwest of Sutter, Sutter County, California; collected by F. H. Holden on 5 April 1912, original number 167; measurements: 303-182-42-13.

Remarks. — *Dipodomys heermanni eximius* (Grinnell, J. Mamm., 2:95, 2 May 1921). Patton, MacArthur, and Yang (J. Mamm., 57:162, 27 February 1976) indicate this subspecies will probably prove referable to *Dipodomys californicus.*

Dipodomys californicus trinitatis Kellogg

Univ. California Publ. Zool., 12(13):366, 27 January 1916.

Type.— Adult male, skin and skull, MVZ 12860; from Helena, 1405 feet altitude, Trinity Mountains region, Trinity County, California; collected by A. M. Alexander on 18 February 1911, original number 1159; measurements: 310-194-46-14.

Remarks.— A synonym of *Dipodomys californicus californicus* according to Grinnell (J. Mamm., 2:97, 2 May 1921). Helena is also called "North Fork," at the junction of the Trinity River and its north fork. Specimens taken from sandy river banks at the bottom of the canyon (of the Trinity River).

Dipodomys deserti aquilus Nader

Proc. Biol. Soc. Washington, 78:52, 21 July 1965.

Type.— Adult male, skin and skull, MVZ 126411; from 1-1/2 miles northwest High Rock Ranch [about 12 miles southeast of Wendel], 4080 feet altitude, Sec. 26, T 28 N, R 17 E, Lassen County, California; collected by C. S. Thaeler on 21 July 1960, original number 1022; measurements: 321-191-54-18, 108.9 g.

Dipodomys heermanni arenae Boulware

Univ. California Publ. Zool., 46(7): 392, 16 September 1943.

Type.— Adult male, skin and skull, MVZ 84840; from C. A. Davis Ranch, 1-3/4 miles north of Lompoc, 400 feet altitude, Santa Barbara County, California; collected by R. M. Bond on 16 February 1939, original number 127; measurements: 290-171-40-16.

Dipodomys heermanni saxatilis Grinnell and Linsdale

Univ. California Publ. Zool., 30(17):453, 15 June 1929.

Type.— Adult female, skin and skull, MVZ 34963; from (mesa near Dale's on north side of) Paine Creek, 700 feet altitude, Tehama County, California; collected by J. and W. F. Grinnell on 27 December 1924, original number (J. Grinnell) 6192; measurements: 300-183-41-12, 57.5 g.

Remarks.— *Dipodomys californicus saxatilis* (Patton, MacArthur, and Yang, J. Mamm., 57:162, 27 February 1976).

Dipodomys jolonensis Grinnell

Proc. Biol. Soc. Washington, 32:203, 31 December 1919.

Type.— Adult male, skin and skull, MVZ 29087; from [valley floor 1 mile southwest of] Jolon [San Antonio River, *ca.* 1000 feet altitude], Monterey County, California; collected by J. Dixon on 18 October 1918, original number 6970; measurements: 310-185-44-15, 82.9 g.

Remarks.— *Dipodomys heermanni jolonensis* (Grinnell, J. Mamm., 2:95, 2 May 1921).

Dipodomys merriami brevinasus Grinnell

J. Mamm., 1(4):179, 24 August 1920.

Type.— Adult male, skin and skull, MVZ 28634; from Hayes Station, near BM 503, 19 miles southwest of Mendota, Fresno County, California; collected by R. Hunt on 30 June 1918, original number 568; measurements: 252-145-36-11, 43.9 g.

Remarks.— *Dipodomys nitratoides brevinasus* (Grinnell, J. Mamm., 2:96, 2 May 1921).

Dipodomys merriami collinus Lidicker

Univ. California Publ. Zool., 67(2):194, 4 August 1960.

Type.— Adult female, skin and skull, MVZ 123455; from 3-1/4 miles south and 2-1/4 miles east of Scissors Crossing, Earthquake Valley, San Diego County, California; collected by W. Z. Lidicker, Jr. on 10 September 1958, original number 1664; measurements: 252-148-39-14, 39.3 g.

Dipodomys merriami vulcani Benson

Proc. Biol. Soc. Washington, 47:181, 2 October 1934.

Type.— Adult male, skin and skull, MVZ 56002; from lower end Toroweap Valley [about 1/2 mile east of Vulcan's Throne], Mohave County, Arizona; collected by A. M. Alexander on 11 November 1932, original number 2064; measurements: 241-138-39-10, 39.5 g.

Dipodomys microps centralis Hall and Dale

Occ. Pap., Mus. Zool., Louisiana State Univ., No. 4:52, 10 November 1939.

Type.— Adult male, skin and skull, MVZ 70817; from 4 miles southeast of Romano, Diamond Valley, Eureka County, Nevada; collected by William B. Richardson on 3 June 1936, original number 1621; measurements: 282-164-43-13, 72.8 g.

Dipodomys microps idahoensis Hall and Dale

Occ. Pap., Mus. Zool., Louisiana State Univ., No. 4:53, 10 November 1939.

Type.— Adult male, skin and skull, MVZ 67568; from 5 miles southeast of Murphy, Owyhee County, Idaho; collected by Howard C. Twining on 26 May 1935, original number 40; measurements: 201-155-43-13.

Remarks.— Original description erroneously lists original number as 39.

Dipodomys microps occidentalis Hall and Dale

Occ. Pap., Mus. Zoo., Louisiana State Univ., No. 4:56, 10 November 1939.

Type.— Adult female, skin and skull, MVZ 64119; from 3 miles south of Schurz, 4100 feet altitude, Mineral County, Nevada; collected by E. R. Hall on 8 July 1934, original number 4158; measurements: 273-160-41-13, 55.8 g.

Dipodomys ordii attenuatus Bryant

Occ. Pap., Mus. Zool., Louisiana State Univ., No. 5:65, 10 November 1939.

Type.— Adult male, skin and skull, MVZ 80429; from mouth of St. Helena Canyon, Big Bend of the Rio Grande River, 2146 feet altitude, [Brewster County] Texas; collected by A. E. Borell on 19 November 1936, original number 5581; measurements: 243-137-37-10.

Dipodomys ordii fetosus Durrant and Hall

Mammalia, 3(1):14, March 1939.

Type.— Adult female, skin and skull, MVZ 48451; from 2 miles north of Panaca, 4800 feet altitude, Lincoln County, Nevada; collected by W. C. Russell on 24 June 1931, original number 1658; measurements: 225-125-41-13.

Dipodomys ordii inaquosus Hall

Proc. Biol. Soc. Washington, 54:58, 20 May 1941.

Type.— Adult male, skin and skull, MVZ 73580; from 11 miles east and 1 mile north of Jungo, 4200 feet altitude, Humboldt County, Nevada; collected by W. C. Russell on 26 July 1936, original number 5026; measurements: 261-142-41-12, 49 g.

Dipodomys ordii priscus Hoffmeister

Proc. Biol. Soc. Washington, 55:167, 31 December 1942.

Type.— Adult female, skin and skull, MVZ 89119; from Kinney Ranch, 7100 feet altitude, 21 miles south of Bittercreek, Sweetwater County, Wyoming; collected by Donald T. Tappe on 16 September 1939, original number 766; measurements: 267-145-41-13, 64.4 g.

Dipodomys ordii terrosus Hoffmeister

Proc. Biol. Soc. Washington, 55:165, 31 December 1942.

Type.— Adult male, skin and skull, MVZ 93477; from the Yellowstone River, 2750 feet altitude, 5 miles west of Forsyth, Rosebud County, Montana; collected by J. R. Alcorn on 2 June 1940, original number 1528; measurements: 266-143-43-12, 71.8 g.

Dipodomys panamintinus caudatus Hall

Mammals of Nevada, p. 409. Univ. California Press, 1 July 1946.

Type.— Adult female, skin and skull, MVZ 80028; from 6 miles south of Granite Well, 3800 feet altitude, Providence Mountains, San Bernardino County, California; collected by F. W. Taber on 18 December 1937, original number 121; measurements: 299-180-43-16, 68.8 g.

Dipodomys sanctiluciae Grinnell

Proc. Biol. Soc. Washington, 32:204, 31 December 1919.

Type.— Adult male, skin and skull, MVZ 29023; from [ridge clothed with digger pine and chaparral, one mile southwest of] Jolon, Monterey County, California; collected by J. Grinnell on 21 October 1918, original number 4950; measurements: 315-190-46-16, 82 g.

Remarks.— *Dipodomys venustus sanctiluciae* (Grinnell, J. Mamm., 2 May 1921).

Dipodomys spectabilis intermedius Nader

Proc. Biol. Soc. Washington, 78:50, 21 July 1965.

Type.— Adult female, skin and skull, MVZ 82782; from 16.7 miles southwest of Bámori, 1900± feet altitude, Sonora, Mexico; collected by Seth B. Benson on 25 April 1938, original number 5301; measurements: 314-180-47-17, 105.7 g.

Perodipus dixoni Grinnell

Univ. California Publ. Zool., 21(2):45, 29 March 1919.

Type.— Adult male, skin and skull, MVZ 26805; from Delhi [near Merced River], Merced County, California; collected by J. Dixon on 23 March 1917, original number 5580; measurements: 280-165-40-14, 72.5 g.

Remarks.— *Dipodomys heermanni dixoni* (Grinnell, J. Mamm., 2:95, 2 May 1921),

Perodipus elephantinus Grinnell

Univ. California Publ. Zool., 21(2):43, 29 March 1919.

Type.— Adult male, skin and skull, MVZ 28511; from 1 mile north of Cook Post Office, 1300 feet altitude, Bear Valley, San Benito County, California; collected by H. G. White on 9 July 1918, original number 2296; measurements: 333-200-48-18, 90.7 g.

Remarks.— *Dipodomys elephantinus* (Grinnell, J. Mamm., 2:96, 2 May 1921).

Perodipus leucogenys Grinnell

Univ. California Publ. Zool., 21(2):46, 29 March 1919.

Type.— Adult male, skin and skull, MVZ 26933; from Pellisier Ranch, 5600 feet altitude, 5 miles north of Benton (Station), Mono County, California; collected by J. Dixon and H. G. White on 20 September 1917, original number (J. Dixon) 6371; measurements: 310-176-48-13, 85.5 g.

Remarks.— *Dipodomys panamintinus leucogenys* (Hall, Mammals of Nevada, p. 407. Univ. California Press, 1 July 1946). Field notes indicate specimen was probably caught by H. G. White and prepared by J. Dixon.

Perodipus mohavensis Grinnell

Univ. California Publ. Zool., 17(14):428, 25 April 1918.

Type.— Adult male, skin and skull, MVZ 26835; from 1/2 mile east of Warren Station, BM 3274 [about 5 miles north of Mohave], Kern County, California; collected by J. Grinnell on 27 March 1917, original number 3942; measurements: 305-178-44-12, 88 g.

Remarks.— *D[ipodomys] panamintinus mohavensis* (Hall, Mammals of Nevada, p. 408. Univ. California Press, 1 July 1946).

Perodipus monoensis Grinnell

Univ. California Publ. Zool., 21(2):46, 29 March 1919.

Type. — Adult female, skin and skull, MVZ 27002; from Pellisier Ranch, 5600 feet altitude, 5 miles north of Benton [Station], Mono County, California; collected by J. Dixon on 21 September 1917, original number 6384; measurments: 245-125-38-12, 48.1 g.

Remarks. — *Dipodomys ordii monoensis* (Grinnell, J. Mamm., 2:96, 2 May 1921).

Perodipus swarthi Grinnell

Univ. California Publ. Zool., 21(2):44, 29 March 1919.

Type. — Adult male, skin and skull, MVZ 14440; from 7 miles southeast of Simmler, Carrizo Plains, San Luis Obispo County, California; collected by H. S. Swarth and W. L. Chandler on 26 May 1911, original number (H. S. Swarth) 9144; measurements: 313-187-45-11.

Remarks. — *Dipodomys heermanni swarthi* (Grinnell, J. Mamm., 2:95, 2 May 1921).

Liomys irroratus acutus Hall and Villa

Univ. Kansas Publ., Mus. Nat. Hist., 1(14):253, 26 July 1948.

Type. — Adult female, skin and skull, MVZ 100171; from 2 miles west of Pátzcuaro, 7700 feet altitude, Michoacán, Mexico; collected by E. R. Hall and J. R. Alcorn on 5 March 1943, original number (J. R. Alcorn) 3837; measurements: 245-122-31-15, 44.8 g.

Remarks. — A synonym of *Liomys irroratus alleni* according to Genoways (Spec. Publ., Mus. Texas Tech. Univ., No. 5:99, 7 December 1973). Original description erroneously indicates specimen was captured on 10 March 1943.

Castoridae

Castor canadensis belugae Taylor

Univ. California Publ. Zool., 12(15):429, 20 March 1916.

Type. — Young adult male, skull only, MVZ 4224; from Beluga River [Cook Inlet region, Kenai Peninsula], Alaska; collected by Jacob Seminoff in 1907, original number 2; no measurements.

Castor canadensis concisor Warren and Hall

J. Mamm., 20(3):358, 14 August 1939.

Type. — Adult [probably male], skull and partial skeleton, MVZ 84521; from Monument Creek, southwest of Monument, El Paso County, Colorado; collected by E. R. Warren, picked up on 20 November 1935, original number 4416; no measurements.

Remarks.— All molars missing; partial skeleton consists of half pelvis, 2 ribs, 1 scapula, and 1 humerus. Skull and other bones found by creek on slope.

Castor canadensis idoneus Jewett and Hall

J. Mamm., 21(1):87, 15 February 1940.

Type.— Adult, sex unknown, skull only, MVZ 86941; from Foley Creek, tributary to Nehalem River, above Nehalem, Tillamook County, Oregon; collected by Edward Leach on 15 December 1914, original number (S. G. Jewett, preparator) 548; measurements: 960-285-170, width of tail 135.

Castor canadensis phaeus Heller

Univ. California Publ. Zool., 5(2):250, 18 February 1909.

Type.— Adult male, study skin and skull, MVZ 209; Pleasant Bay, Admiralty Island, Alaska; collected by A. Hasselborg on 16 May 1907, original number (A. M. Alexander) 27; measurements: 1030-485-175.

Castor canadensis sagittatus Benson

J. Mamm., 14(4):320, 13 November 1933.

Type.— Young adult female, flat skin and complete skeleton, MVZ 43906; from Indianpoint Creek, 3200 feet altitude [16 miles northeast of Barkerville], British Columbia; collected by T. T. and E. B. McCabe on 2 October 1929, original number 500; measurements: 967-167-36.

Castor canadensis taylori Davis

The Recent mammals of Idaho, p. 273, The Caxton Printers, Ltd., 5 April 1939.

Type.— Adult female, flat skin and skull, MVZ 67588; from [Big Wood River, near] Bellvue [=Bellevue], Blaine County, Idaho; collected by J. M. Wright on 10 April 1935, original number (W. B. Davis) 1137; no measurements.

Remarks.— This specimen was in an express shipment received at MVZ on 18 April 1935. E. R. Hall and W. B. Davis associated the poorly labeled skins with skulls, and determined the makeup of this specimen on 14 February 1936.

Castor subauratus Taylor

Univ. California Publ. Zool., 10(7):167, 21 May 1912.

Type.— Adult female, flat skin with complete skeleton, MVZ 12654; from Grayson, San Joaquin River [Stanislaus County], California; collected by Jack Doyle on 21 February 1911, no original number; measurements: 1171-320-196-31, 39-1/2 lbs (with entrails removed).

Remarks.— *Castor canadensis subauratus* (Grinnell, Univ. California Publ. Zool., 40:166, 26 September 1933). Notes by A. C. Ziegler and S. B. Benson indicate that from the accession card and what appears to be Grinnell's original tag on the skin, the type locality should be "5 miles north of Grayson(s) on San Joaquin River." Doyle's letters, however, never state the exact locality, so the source of Grinnell's "5 miles north of Grayson(s)" is not known.

Cricetidae

Nyctomys sumichrasti florencei Goldman

J. Washington Acad. Sci., 27(10):421, 15 October 1937.

Type.— Adult female, skin with complete skeleton, MVZ 131420; from Barra de Santiago, sea level, Department Ahuachapan, El Salvador; collected by R. A. Stirton on 6 April 1927, original number (collection of Donald R. Dickey) 12765; measurements: 238-127-21-14.

Remarks.— Caught on leaning tree in swamp forest with fruit pulp in stomach.

Reithrodontomys burti Benson

Proc. Biol. Soc. Washington, 52:147, 11 October 1939.

Type.— Adult male, skin and skull plus baculum, MVZ 83001; from Rancho de Costa Rica, 270 feet altitude, Rio Sonora, Sonora, Mexico; collected by M. Delgadillo on 3 May 1938, original number (Seth B. Benson) 5400; measurements: 128-61-16-15, 9.9 g.

Remarks.— Specimen label indicates the type was taken in wheat stubble.

Reithrodontomys fulvescens canus Benson

Proc. Biol. Soc. Washington, 52:149, 11 October 1939.

Type.— Adult male, skin and skull, MVZ 76664; from 5 miles southeast of Chihuahua, 5250 feet altitude, Chihuahua, Mexico; collected by Seth B. Benson and M. Delgadillo on 20 May 1937, original number (S. B. Benson) 4446; measurements: 170-94-20-16, 13.5 g.

Reithrodontomys halicoetes Dixon

Univ. California Publ. Zool., 5(4):271, 14 August 1909.

Type.— Adult male, skin and skull, MVZ 7146; from [three miles south of] Petaluma, Sonoma County, California; collected by J. Dixon on 28 March 1908, original number 179; measurements: 155-85-16.

Remarks.— *Reithrodontomys raviventris halicoetes* (Howell, N. Amer. Fauna, 36:42, 5 June 1914).

Reithrodontomys megalotis distichlis von Bloeker

Proc. Biol. Soc. Washington, 50:155, 10 September 1937.

Type.— Adult male, skin and skull, MVZ 74683; from [salt-marsh at the] mouth, Salinas River, Monterey County, California; collected by Jack C. von Bloeker, Jr. on 10 June 1936, original number 6271; measurements:155-85-17-14.

Reithrodontomys megalotis santacruzae Pearson

J. Mamm., 32(3):366, 23 August 1951.

Type.— Adult male, skin and skull, MVZ 113604; from Prisoner's Harbor, Santa Cruz Island, Santa Barbara County, California; collected by Alden H. Miller (original number 7579) on 7 March 1950, prepared by O. P. Pearson, original number 1938; measurements: 156-88-18-15, 12.3 g.

Reithrodontomys mexicanus orinus Hooper

Proc. Biol. Soc. Washington, 62:169, 16 November 1949.

Type. — Adult male, skin and skull, MVZ 98459; from Hacienda Chilata, 2000 feet altitude (near the summit of the Balsam Range, about 12 miles south of Sonsonate), Department Sonsonate, El Salvador; collected by M. Hildebrand on 12 May 1942, original number 1465; measurements: 184-105-19-15.

Reithrodontomys raviventris Dixon

Proc. Biol. Soc. Washington, 21:197, 20 October 1908.

Type. — Male, skin and skull, MVZ 475; from [salt marsh near] Redwood City [San Mateo County], California; collected by C. Littlejohn on 15 January 1908, original number 134; measurements: 130-69-17.

Peromyscus californicus benitoensis Grinnell and Orr

J. Mamm., 15(3):216, 10 August 1934.

Type. — Adult female, skin with complete skeleton, MVZ 28197; from near Cook Post Office, 1300 feet altitude, Bear Valley, San Benito County, California; collected by R. Hunt on 12 July 1918, original number 634; measurements: 260-132-28-25, 48.5 g.

Peromyscus californicus mariposae Grinnell and Orr

J. Mamm., 15(3):217, 10 August 1934.

Type. — Adult male, skin and skull, MVZ 21703; from El Portal, 2500 feet altitude, Mariposa County, California; collected by Walter P. Taylor on 1 December 1914, original number 7159; measurements: 248-136-27-22.

Peromyscus crinitus delgadilli Benson

Proc. Biol. Soc. Washington, 53:1, 16 February 1940.

Type. — Adult male, skin and skull plus baculum, MVZ 83042; from 2 miles south of Crater Elegante, 900 ± feet altitude [Sierra del Pinacate], 34 miles west Sonoita, Sonora, Mexico; collected by M. Delgadillo on 28 March 1938, original number (Seth B. Benson) 4963; measurements: 179-104-19-20, 11.7 g.

Peromyscus crinitus rupicolus Benson

Proc. Biol. Soc. Washington, 53:2, 16 February 1940.

Type. — Adult female, skin and skull, MVZ 83034; from Paso MacDougal, 850 ± feet elevation, Pinacate region [east end of Sierra Hornaday], Sonora, Mexico; collected by Seth B. Benson on 25 March 1938, original number 4944; measurements: 185-108-19-21, 10.9 g.

Remarks. — A synonym of *Peromyscus crinitus disparilis* according to Hall and Hoffmeister (J. Mamm., 23:62, 14 February 1942).

Peromyscus crinitus scopulorum Benson

Proc. Biol. Soc. Washington, 53:2, 16 February 1940.

Type.— Adult male, skin and skull plus baculum, MVZ 83045; from Cerro La Cholla, 50 ± feet altitude, 6 miles west-northwest of Punta Peñasca, Sonora, Mexico; collected by Seth B. Benson on 3 April 1938, original number 5024; measurements: 197-115-19-20, 17.0 g.

Remarks.— A synonym of *Peromyscus crinitus disparilis* according to Hall and Hoffmeister (J. Mamm., 23:62, 14 February 1942).

Peromyscus eremicus cinereus Hall

Proc. Biol. Soc. Washington, 44:87, 29 June 1931.

Type.— Adult female, skin and skull, MVZ 43025; from southwest end of San José Island [25° N latitude], Baja California, Mexico; collected by Chester C. Lamb on 5 March 1929, original number 10576; measurements: 185-105-20-14, 16.6 g.

Peromyscus leucopus caudatus Smith

Proc. Biol. Soc. Washington, 52:157, 11 October 1939.

Type.— Adult male, skin and skull, MVZ 84535; from Wolfville (Kings County), Nova Scotia; collected by Ronald W. Smith on 11 November 1937, original number 1502; measurements: 187-99-21.

Peromyscus maniculatus angustus Hall

Univ. California Publ. Zool., 38(12):423, 8 November 1932.

Type.— Adult female, skin and skull, MVZ 12269; from [Beaver Creek, 15 miles northwest of Alberni], Alberni Valley, Vancouver Island, British Columbia; collected by L. Kellogg on 25 June 1910, original number 958; measurements: 179-83-20.

Peromyscus maniculatus georgiensis Hall

Amer. Natur., 72(742):455, 10 September 1938.

Type.— Subadult female, skin and skull, MVZ 70400; from Vanada, Texada Island [Strait of Georgia], British Columbia; collected by R. A. Cumming on 4 May 1936, original number 1518; measurements: 170-70-23.

Peromyscus maniculatus serratus Davis

Recent mammals of Idaho, p. 290, The Caxton Printers, Ltd., 5 April 1939.

Type.— Adult female, skin and skull, MVZ 72330; from Mill Creek, 8370 feet altitude, 14 miles west-southwest of Challis, Custer County, Idaho; collected by W. B. Davis on 27 July 1936, original number 2290; measurements: 201-95-24-22.

Peromyscus nasutus griseus Benson

Univ. California Publ. Zool., 38(5):338, 14 April 1932.

Type.— Adult male, skin and skull, MVZ 50819; from the Malpais lava beds, [3-1/2 miles west of] Carrizozo, 5150 feet altitude, Lincoln County, New Mexico;

collected by L. Kellogg on 26 September 1931, original number 1149; measurements: 201-102-23-18, 25 g.

Peromyscus sitkensis oceanicus Cowan

Univ. California Publ. Zool., 40(13):432, 14 November 1935.

Type.— Adult male, skin and skull, MVZ 20890; from Forrester Island, Alaska; collected by H. and R. W. Heath on 4 July 1913, original number 26; measurements: 216-101-27-16.

Peromyscus truei chlorus Hoffmeister

Proc. Biol. Soc. Washington, 54:131, 30 September 1941.

Type.— Young adult female, skin and skull, MVZ 77194; from Lost Horse Mine, 5500 feet altitude [=69 miles east of Riverside], southern end of the Little San Bernardino Mountains, Riverside County, California; collected by Robert D. Moore on 9 March 1929, original number 163; measurements: 196-104-21-24.

Peromyscus truei nevadensis Hall and Hoffmeister

Univ. California Publ. Zool., 42(8):401, 30 April 1940.

Type.— Adult female, skin and skull, MVZ 68479; from Pilot Peak, 6000 feet altitude, 1/2 mile west of Debbs Creek, Elko County, Nevada; collected by A. E. Peterson on 20 July 1935, original number 144; measurements: 212-97-24-27, 38.5 g.

Peromyscus truei sequoiensis Hoffmeister

Proc. Biol. Soc. Washington, 54:129, 30 September 1941.

Type.— Adult male, skin and skull, MVZ 20842; from 1 mile west of Guerneville, Sonoma County, California; collected by H. W. Grinnell on 16 June 1913, original number 109; measurements: 222-122-26.

Graomys pearsoni Myers

Occ. Pap., Mus. Zool., Univ. Michigan, No. 676:1, 18 March 1977.

Type.— Adult male, skin and skull, MVZ 145276; from 410 kilometers northwest Villa Hayes by road, Departamento Boquerón, Paraguay; collected by P. Myers on 24 September 1973, original number 1161; measurements: 236-124-25-20, 50 g.

Remarks.— *Andalgalomys pearsoni* (Williams and Mares, Ann. Carnegie Mus., 47(9):210, 28 June 1978).

Phyllotis caprinus Pearson

Univ. California Publ. Zool., 56(4):435, 8 December 1958.

Type.— Adult female, skin with complete skeleton, MVZ 120210; from Tilcara, 8000 feet altitude, Jujuy, Argentina; collected by O. P. Pearson on 1 October 1955, original number 4256; measurements: 252-133-27-26.

Remarks.— *Phyllotis (Phyllotis) caprinus* (Pearson and Patton, J. Mamm., 57:339, 20 May 1976).

Sigmodon hispiduas (sic) atratus Hall

Proc. Biol. Soc. Washington, 62:149, 23 August 1949.

Type. — Subadult male, skin and skull, MVZ 100628; from 6-1/2 miles west of Zamora, 5950 feet altitude, Michoacán, Mexico; collected by E. R. Hall on 27 March 1943, original number 6009; measurements: 228-112-30-18, 53.2 g.

Remarks. — A synonym of *Sigmodon hispidus inexoratus* according to Russell (Proc. Biol. Soc. Washington, 65:82, 25 April 1952).

Sigmodon ochrognathus montanus Benson

Proc. Biol. Soc. Washington, 53:157, 19 December 1940.

Type. — Adult male, skin and skull plus baculum, MVZ 92287; from Peterson's Ranch [also called Sylvania], 6100 feet altitude, Huachuca Mountains, 2 miles north of Sunnyside, Cochise County, Arizona; collected by Seth B. Benson on 15 March 1940, original number 6649; measurements: 233-100-28-18, 78.1 g.

Neotoma albigula brevicauda Durrant

J. Mamm., 15(1):65, 15 February 1934.

Type. — Adult male, skin and skull, MVZ 57347; from Castle Valley, 4500 feet altitude, about 15 miles northeast of Moab, Grand County, Utah; collected by S. D. Durrant on 13 May 1933, original number 47; measurements: 313-122-30-29.

Neotoma cinerea alticola Hooper

Univ. California Publ. Zool., 42(9):409, 17 May 1940.

Type. — Adult female, skin and skull, MVZ 11152; from Parker Creek [=Shields Creek, U. S. Forest Service map, edition of 1932], 5500 feet altitude, Warner Mountains, Modoc County, California; collected by N. B. Stern on 22 June 1910, original number 153; measurements: 389-178-44-26.

Neotoma cinerea pulla Hooper

Univ. California Publ. Zool., 42(9):411, 17 May 1940.

Type. — Adult female, skin and skull, MVZ 57077; from Kohnenberger's Ranch, 3200 feet altitude, South Fork Mountain, Trinity County, California; collected by Chester C. Lamb on 25 June 1932, original number 16758; measurements: 396-193-41-29, 298.3 g.

Neotoma fuscipes bullatior Hooper

Univ. California Publ. Zool., 42(4):225, 1 March 1938.

Type. — Adult female, skin and skull, MVZ 28900; from 2 miles south of San Miguel, 620 feet altitude, San Luis Obispo County, California; collected by R. Hunt on 9 November 1918, original number 742; measurements: 455-221-43-37, 345.5 g.

Remarks. — Left parietal region of skull damaged.

Neotoma fuscipes luciana Hooper

Univ. California Publ. Zool., 42(4):229, 1 March 1938.

Type. — Adult female, skin and skull, MVZ 29321; from beneath live oak tree, Seaside, Monterey County, California; collected by H. G. White on 17 December 1918, original number 2884; measurements: 408-198-40-28, 269 g.

Neotoma fuscipes martirensis Orr

Proc. Biol. Soc. Washington, 47:110, 13 June 1934.

Type. — Adult male, skin and skull, MVZ 35850; from Valladares, 2700 feet altitude [Sierra San Pedro Mártir], Baja California, Mexico; collected by A. E. Borell and C. C. Lamb on 15 April 1925, original number (A. E. Borell) 1419; measurements: 390-180-38-30, 260 g.

Neotoma fuscipes perplexa Hooper

Univ. California Publ. Zool., 42(4):224, 1 March 1938.

Type. — Adult female, skin and skull, MVZ 14079; from Sweeney's Ranch, 22 miles south of Los Banos, Merced County, California; collected by W. L. Chandler and C. H. Richardson on 30 March 1911, original number (C. H. Richardson) 65; measurements: 413-206-40.

Neotoma fuscipes riparia Hooper

Univ. California Publ. Zool., 42(4):223, 1 March 1938.

Type. — Adult male, skin and skull, MVZ 55130; from Kincaid's Ranch, 2 miles northeast of Vernalis, Stanislaus County, California; collected by Robert T. Orr on 4 November 1932, original number 861; measurements: 448-220-45-36.

Neotoma lepida egressa Orr

Proc. Biol. Soc. Washington, 47:109, 13 June 1934.

Type. — Adult male, skin and skull, MVZ 50142; from 1 mile east of El Rosario, 200 feet altitude, Baja California, Mexico; collected by Chester C. Lamb on 26 December 1930, original number 13344; measurements: 340-152-35-34., 173 g.

Neotoma lepida flava Benson

Occ. Pap. Mus. Zool., Univ. Michigan, No. 317:7, 1 July 1935.

Type. — Adult female, skin and skull, MVZ 62657; from Tinajas Altas, 1150 feet altitude, Yuma County, Arizona; collected by M. M. Erickson on 17 January 1934, original number 203; measurements: 293-147-30-31, 87.8 g.

Neotoma lepida grinnelli Hall

Univ. California Publ. Zool., 46(5):369, 3 July 1942.

Type. — Adult male, skin and skull, MVZ 10438; from Colorado River, 20 miles [by river, but about 12-1/2 miles north by air-line] above Picacho, Imperial

County, California; collected by Frank Stephens on 13 April 1910, original number 2694; measurements: 295-138-30-30.

Neotoma lepida petricola von Bloeker

Proc. Biol. Soc. Washington, 51:203, 23 December 1938.

Type.— Adult female, skin and skull, MVZ 30202; from Abbott Ranch [670 feet altitude], Arroyo Seco, Monterey County, California; collected by H. G. White on 19 July 1919, original number 3525; measurements: 352-175-32-28, 174 g.

Remarks.— Taken in sand beneath bowlder (sic).

Neotoma nevadensis Taylor

Univ. California Publ. Zool., 5(6):289, 12 February 1910.

Type.— Adult female, skin and skull, MVZ 8282; from Virgin Valley [4800 feet altitude, Humboldt County], Nevada; collected by A. M. Alexander on 17 May 1909, original number 23; measurements: 260-115-30.

Remarks.— *Neotoma lepida nevadensis* (Hall, Mammals of Nevada, p. 530, Univ. California Press, 1 July 1946).

Clethrionomys albiventer Hall and Gilmore

Univ. California Publ. Zool., 38(9):398, 17 September 1932.

Type.— Adult female, skin and skull, MVZ 51221; from Sevoonga [2 miles east of North Cape], St. Lawrence Island, Bering Sea, Alaska; collected by R. M. Gilmore on 27 June 1931, original number 1660; measurements: 153-34-21-13.

Remarks.— *C[lethrionomys] rutilus albiventer* (Rausch, J. Parasit., 38:416, October 1952).

Clethrionomys dawsoni glacialis Orr

J. Mamm., 26(1):71, 23 February 1945.

Type.— Adult male, skin and skull, MVZ 388; from [Bartlett Cove] Glacier Bay, Alaska; collected by Frank Stephens on 11 July 1907, original number 158; measurements: 146-39-21-13.

Remarks.— *Clethrionomys rutilus glacialis* (Rausch, J. Washington Acad. Sci., 40:135, 15 April 1950).

Clethrionomys gapperi pallescens Hall and Cockrum

Univ. Kansas Publ., Mus. Nat. Hist., 5(23):302, 17 November 1952.

Type.— Adult male, skin and skull, MVZ 86721; from Wolfville [Kings County], Nova Scotia; collected by R. W. Smith on 7 May 1938, original number 1615; measurements: 146.5-47.5-19.2.

Remarks.— A new name for *Clethrionomys gapperi rufescens* (Smith, Amer. Midland Natur., 24(1):233, 31 July 1940), which was preoccupied.

Evotomys dawsoni insularis Heller

Univ. California Publ. Zool., 5(11):339, 5 March 1910.

Type.— Adult male, skin and skull, MVZ 557; from west side of Canoe Passage, Hawkins Island, Prince William Sound, Alaska; collected by E. Heller on 20 June 1908, original number 79; measurements: 143-32-19.5.

Remarks.— *Clethrionomys rutilus insularis* (Rausch, J. Washington Acad. Sci., 40:135, 21 April 1950).

Evotomys phaeus Swarth

Univ. California Publ. Zool., 7(2):127, 12 January 1911.

Type.— Adult male, skin and skull, MVZ 8742; from [head of Marten Arm] Boca de Quadra, Alaska; collected by H. S. Swarth on 13 June 1909, original number 7647; measurements: 156-58-19.

Remarks.— *Clethrionomys gapperi phaeus* (Hall and Cockrum, Univ. Kansas Publ., Mus. Nat Hist., 5:302, 17 November 1952).

Microtus admiraltiae Heller

Univ. California Publ. Zool., 5(2):256, 18 February 1909.

Type.— Adult female, skin and skull, MVZ 118; from Windfall Harbor, Admiralty Island, Alaska; collected by J. Dixon on 4 May 1907, original number 25; measurements: 158-48-21.

Remarks.— *Microtus pennsylvanicus admiraltiae* (Swarth, Proc. Biol. Soc. Washington, 46:208, 26 October 1933).

Microtus californicus aestuarinus Kellogg

Univ. California Publ. Zool., 21(1):15, 28 December 1918.

Type.— Adult female, skin and skull, MVZ 18699; from Grizzly Island, Solano County, California; collected by A. M. Alexander on 8 September 1912, original number 1942; measurements: 205-62-25.

Microtus californicus eximius Kellogg

Univ. California Publ. Zool., 21(1):12, 28 December 1918.

Type.— Adult male, skin and skull, MVZ 20098; from Lierly's Ranch, 2340 feet altitude, 4 miles south of Mount Sanhedrin, Menocino County, California; collected by C. L. Camp on 22 August 1913, original number 1299; measurements: 183-58-22-12.

Microtus californicus halophilus von Bloeker

Proc. Biol. Soc. Washington, 50:156, 10 September 1937.

Type.— Adult female, skin and skull, MVZ 74684; from Moss Landing, Monterey County, California; collected by Jack C. von Bloeker, Jr. on 11 August 1936, original number 7079; measurements: 173-48-21-15.

Microtus californicus kernensis Kellogg

Univ. California Publ. Zool., 21(1):26, 28 December 1918.

Type. — Adult female, skin and skull, MVZ 15779; from Fay Creek, 4100 feet altitude, 6 miles north of Weldon, Kern County, California; collected by Tracy I. Storer on 14 July 1911, original number 219; measurements: 179-64-23.

Microtus californicus mariposae Kellogg

Univ. California Publ. Zool., 21(1):19, 28 December 1918.

Type. — Adult male, skin and skull, MVZ 21724; from flat, 1-3/4 miles west of El Portal, 1800 feet altitude, Mariposa County, California; collected by Walter P. Taylor on 2 December 1914, original number 7166; measurements: 193-56-24.

Microtus californicus mohavensis Kellogg

Univ. California Publ. Zool., 21(1):29, 28 December 1918.

Type. — Adult male, skin and skull, MVZ 5987; from Victorville [2700 feet altitude, San Bernardino County], California; collected by J. Grinnell and W. P. Taylor on 6 April 1907, original number (J. Grinnell) 2036; measurements: 214-63-25.

Microtus californicus paludicola Hatfield

J. Mamm., 16(4):316, 15 November 1935.

Type. — Adult male, skin and skull, MVZ 12718; from Melrose Marsh, Alameda County, California; collected by A. M. Alexander on 15 December 1910, original number 1108; measurements: 157-39-20.5.

Remarks. — A synonym of *M[icrotus] c[alifornicus] californicus* according to Thaeler (Univ. California Publ. Zool., 60(2):81, 28 November 1961).

Microtus californicus perplexabilis Grinnell

J. Mamm., 7(3):223, 9 August 1926.

Type. — Adult male, skin and skull, MVZ 35882; from La Grulla, 7500 feet altitude, Sierra San Pedro Mártir, Baja California; collected by A. E. Borell on 13 May 1925, original number 1642; measurements: 195-57-26-11, 66 g.

Remarks. — A synonym of *Microtus californicus huperuthrus* according to Hall and Cockrum (Univ. Kansas Publ., Mus. Nat. Hist., 5:420, 15 January 1953). Original description erroneously gives altitude as 7000 feet.

Microtus californicus sanctidiegi Kellogg

Proc. Biol. Soc. Washington, 35:78, 20 March 1922.

Type. — Adult female, skin and skull, MVZ 19119; from Escondido [640 feet altitude], San Diego County, California; collected by J. Dixon on 7 February 1913, original symbol = Δ; measurements: 194-51-23.

Remarks. — A new name for *Microtus californicus neglectus* (Kellogg, Univ. California Publ. Zool., 21(1):31, 28 December 1918), which was preoccupied.

Microtus californicus sanpabloensis Thaeler

Univ. California Publ. Zool., 60(2):81, 28 November 1961.

Type.— Adult female, skin and skull, MVZ 123743; from San Pablo Creek (saltmarsh), Richmond, Contra Costa County, California; collected by C. S. Thaeler, Jr. on 23 December 1958, original number 748; measurements: 170-48-20-14, 50.8 g.

Microtus coronarius Swarth

Univ. California Publ. Zool., 7(2):131, 12 January 1911.

Type.— Adult female, skin and skull, MVZ 8721; from Egg Harbor, Coronation Island, Alaska; collected by H. S. Swarth on 16 May 1909, original number 7487; measurements: 215-84-25.

Remarks.— Goldman (J. Mamm., 19:491, 14 November 1938) suggests that future studies may show that *Microtus coronarius* should be regarded as a subspecies of *Microtus longicaudus*.

Microtus innuitus punukensis Hall and Gilmore

Univ. California Publ. Zool., 38(9):399, 17 September 1932.

Type.— Adult female, skin and skull, MVZ 51392; from Big Punuk Island, near east end of St. Lawrence Island, Bering Sea, Alaska; collected by O. W. Geist on 19 August 1931, original number 4 (R. M. Gilmore 2330); measurements: 196-40.5-22.5-17.5.

Remarks.— M[icrotus] oec[onomus] punukensis (Zimmerman, Arch. Naturg., Berlin, n.F., vol. 11, no. 2, p. 188, 10 September 1942).

Microtus (Lagurus) intermedius Taylor

Univ. California Publ. Zool., 7(7):253, 24 June 1911.

Type.— Adult male, skin and skull, MVZ 7973; from head of Big Creek, 8000 feet altitude, Pine Forest Mountains, Humboldt County, Nevada; collected by C. H. Richardson, Jr. and W. P. Taylor on 1 July 1909, original number 3082; measurements: 120-24-18.

Remarks.— *Lagurus curtatus intermedius* (Borell and Ellis, J. Mamm., 15:35, 15 February 1934). Occipital region of skull damaged.

Microtus mexicanus fundatus Hall

Univ. Kansas Publ., Mus. Nat. Hist., 1(21):425, 24 December 1948.

Type.— Adult male, skin and skull, MVZ 100637; from 3-1/2 miles south of Pátzcuaro, 7900 feet altitude, Michoacán, Mexico; collected by E. R. Hall on 9 March 1943, original number 5882; measurements: 154-36-21-13, 41 g.

Remarks.— Original description notes specimens of this taxon were "taken in the dry season, trapped mostly in runways beneath a dense growth of grass underneath a rail fence."

Microtus mexicanus navaho Benson

Proc. Biol. Soc. Washington, 47:49, 9 February 1934.

Type.— Adult male, skin and skull, MVZ 58817; from Soldier Spring, 8800 feet altitude [east slope of] Navajo Mountain, San Juan County, Utah; collected by Seth B. Benson on 17 June 1933, original number 2155; measurements: 136-33-18-13, 31.8 g.

Remarks.— Taken in *Ceanothus.*

Microtus montanus amosus Hall and Hayward

Great Basin Natur., 2(2):105, 20 July 1941.

Type.— Adult female, skin and skull, MVZ 95272; from Torrey, Wayne County, Utah; collected by James W. Bee on 18 June 1938, original number 705; measurements: 179-54-20.5, 74 g.

Microtus montanus fucosus Hall

Univ. California Publ. Zool., 40(12):421, 25 October 1935.

Type.— Adult male, skin and skull, MVZ 53248; from Hiko, 4000 feet altitude, Pahranagat Valley, Lincoln County, Nevada; collected by W. C. Russell on 21 May 1932, original number 2116; measurements: 179-46-24-13, 53.6 g.

Microtus montanus fusus Hall

Proc. Biol. Soc. Washington, 51:131, 23 August 1938.

Type.— Adult male, skin and skull, MVZ 61281; from 2-1/2 miles east of summit, Cochetopa Pass, Saguache County, Colorado; collected by A. M. Alexander on 21 September 1933, original number 2568; measurements: 156-40-19-10, 40 g.

Microtus montanus kincaidi Dalquest

Proc. Biol. Soc. Washington, 54:145, 30 September 1941.

Type.— Adult female, skin and skull, MVZ 95084; from The Potholes [ten miles south] of Moses Lake [Neppel], Grant County, Washington; collected by Walter W. Dalquest on 24 March 1940, original number 1748; measurements: 166-44-20.5-14.

Remarks.— *Microtus pennsylvanicus kincaidi* (Dalquest, Univ. Kansas Publ., Mus. Nat. Hist., 2:347, 9 April 1948).

Microtus montanus micropus Hall

Univ. California Publ. Zool., 40(12):417, 25 October 1935.

Type.— Adult male, skin and skull, MVZ 46282; from Cleveland Ranch, 6000 feet altitude, Spring Valley, White Pine County, Nevada; collected by W. C. Russell on 30 July 1930, original number 721; measurements: 177-43-22-14, 64 g.

Microtus montanus nexus Hall and Hayward

Great Basin Natur., 2(2):106, 20 July 1941.

Type. — Adult female, skin and skull, MVZ 95271; from West Canyon, Oquirrh Range, Utah County, Utah; collected by James W. Bee on 3 August 1939, original number 19-8-3-39; measurements: 155-41.2-20, 50 g.

Remarks. — Skin prepared as flat body.

Microtus montanus undosus Hall

Univ. California Publ. Zool., 40(12):420, 25 October 1935.

Type. — Adult female, skin and skull, MVZ 5828; from Lovelock [Pershing County], Nevada; collected by J. M. Willard on 10 April 1901, original number 9; measurements: 12.3-5.4-5.2-3.6 (sic).

Microtus montanus yosemite Grinnell

Proc. Biol. Soc. Washington, 27:207, 31 October 1914.

Type. — Adult female, skin and skull, MVZ 12978; from Yosemite Valley, 4000 feet altitude, Mariposa County, California; collected by J. and H. W. Grinnell on 27 May 1911, original number (J. Grinnell) 675; measurements: 166-51-20.

Remarks. — Grinnell began a new series of field numbers on 7 November 1910.

Microtus mordax halli Hayman and Holt

In Ellerman, The families and genera of living rodents, British Mus. Nat. Hist., 2:603, 21 March 1941.

Type. — Adult female, skin and skull, MVZ 40138; from Godman (=Goodman) Spring, 5700 feet altitude, Blue Mountains [Columbia County], Washington; collected by S. H. Lyman on 1 September 1927, original number H 23; measurements: 179-63-19.5.

Remarks. — A new name for *Microtus mordax angustus* (Hall, Univ. California Publ. Zool., 37(1):13, 10 April 1931), which was preoccupied. *Microtus longicaudus halli* (Dalquest, Univ. Kansas Publ., Mus. Nat. Hist., 2:353, 9 April 1948).

Microtus mordax latus Hall

Univ. California Publ. Zool., 37(1):12, 10 April 1931.

Type. — Adult female, skin and skull, MVZ 45846; from Wisconsin Creek, 8500 feet altitude, Toyabe Mountains, Nye County, Nevada; collected by Chester C. Lamb on 28 May 1930, original number 12409; measurements: 184-52-25-12, 58 g.

Remarks. — *Microtus longicaudus latus* (Goldman, J. Mamm., 19:491, 14 November 1938).

Microtus mordax littoralis Swarth

Proc. Biol. Soc. Washington, 46:209, 26 October 1933.

Type. — Adult male, skin and skull, MVZ 8642; from Shakan, Prince of Wales Island, Alaska; collected by H. S. Swarth on 14 May 1909, original number 7463; measurements: 189-72-22.

Remarks. — *Microtus longicaudus littoralis* (Goldman, J. Mamm., 19:491, 14 November 1938). Hall and Kelson (The mammals of North America, 2:738, 31 March 1959) comment that "Shakan is on Kosciusko Island and Swarth probably took his specimen on Prince of Wales Island, opposite Shakan."

Microtus mordax sierrae Kellogg

Univ. California Publ. Zool., 21(8):288, 18 April 1922.

Type.— Adult male, skin and skull, MVZ 22437; from Tuolumne Meadows, 8600 feet altitude, Yosemite National Park, Tuolumne County, California; collected by C. L. Camp on 7 July 1915, original number 2159; measurements: 194-67-22-16, 75.8 g.

Remarks.— *Microtus longicaudus sierrae* (Goldman, J. Mamm., 19:491, 14 November 1938).

Microtus pennsylvanicus funebris Dale

J. Mamm., 21(3):338, 13 August 1940.

Type.— Adult male, skin and skull, MVZ 77482; from Coldstream, 1450 feet altitude, 3-1/2 miles southeast of Vernon, British Columbia; collected by T. P. Maslin on 2 August 1937; original number 741; measurements: 172-52-20-14.

Remarks.— A synonym of *Microtus pennsylvanicus drummondii* according to Cowan and Guiguet (Handbook No. 11:219, British Columbia Prov. Mus., 15 July 1956).

Microtus pennsylvanicus rubidus Dale

J. Mamm., 21(3):339, 13 August 1940.

Type.— Adult male, skin and skull, MVZ 30758; from Sawmill Lake, near Telegraph Creek (Stikine River, Cassiar Electric District), British Columbia; collected by J. Dixon on 12 June 1919, original number 7361; measurements: 151-36-19-10, 35.6 g.

Remarks.— A synonym of *Microtus pennsylvanicus drummondii* according to Cowan and Guiguet (Handbook No. 11:219, British Columbia Prov. Mus., 15 July 1956).

Microtus townsendii cummingi Hall

Murrelet, 17(1):15, 28 March 1936.

Type.— Adult male, skin and skull, MVZ 68836; from Bowen Island [Howe Sound], British Columbia; collected by R. A. Cumming on 5 October 1935, original number 1107; measurements: 181-52-25.

Microtus townsendii pugeti Dalquest

Murrelet, 21(1):7, 30 April 1940.

Type.— Adult male, skin and skull, MVZ 83427; from Neck Point, [northwest corner of] Shaw Island, San Juan County, Washington; collected by D. H. Johnson on 10 July 1938, original number 3439; measurements: 190-53-23-15.

Zapodidae

Zapus princeps cinereus Hall

Univ. California Publ. Zool., 37(1):7, 10 April 1931.

Type.— Adult female, skin and skull, MVZ 45422; from Pine Canyon, 6600 feet altitude, 17 miles northwest of Kelton, Raft River Mountains, Box Elder County, Utah; collected by A. M. Alexander on 14 July 1930, original number 689; measurements: 235-135-32-12.

Zapus princeps curtatus Hall

Univ. California Publ. Zool., 37(1):7, 10 April 1931.

Type.— Adult female, skin and skull, MVZ 7991; from head of Big Creek, 8000 feet altitude, Pine Forest Mountains [Humboldt County], Nevada; collected by W. P. Taylor and C. H. Richardson on 30 June 1909, original number (W. P. Taylor) 777; measurements: 235-141-33.

Zapus princeps idahoensis Davis

J. Mamm., 15(3):221, 10 August 1934.

Type.— Adult male, skin and skull, MVZ 54845; from 5 miles east of Warm Lake, 7000 feet altitude, Valley County, Idaho; collected by Robert T. Orr on 9 July 1932, original number 660; measurements: 231-139-30-15.5, 22.7 g.

Zapus princeps palatinus Hall

Univ. California Publ. Zool., 37(1):8, 10 April 1931.

Type.— Adult male, skin and skull, MVZ 45871; from Wisconsin Creek, 7800 feet altitude, Toyabe Mountains, Nye County, Nevada; collected by J. Linsdale on 26 May 1930, original number 3191; measurements: 226-135-34-12, 26 g.

Remarks.— A synonym of *Zapus princeps oregonus* according to Krutzsch (Univ. Kansas Publ., Mus. Nat. Hist., 7:409, 21 April 1954).

Zapus trinotatus eureka Howell

Univ. California Publ. Zool., 21(5):229, 20 May 1920.

Type.— Adult female, skin and skull, MVZ 11703; from Fair Oaks (2), Humboldt County, California; collected by J. Dixon on 27 August 1910, original number 1743; measurements: 237-148-32.

Remarks.— Dixon's field notes indicate that Fair Oaks (2) is located *ca.* 6 miles northwest of Fair Oaks Post Office, at the top of Kneeland Grade, just above Hay's old sawmill, 1500-1800 feet altitude.

Erethizontidae

Erethizon epixanthum bruneri Swenk

Univ. Stud., Lincoln, Nebraska, 16 (1-2): 117, 21 November 1916.

Type. — Adult male, study skin and skull, MVZ 31863; from 3 miles east of Mitchell, Scottsbluff County, Nebraska; collected by J. E. Dorothy on 26 September 1915, original number (Myron H. Swenk) 305; measurements: 797-202-117, 22-1/2 lbs.

Remarks. — *Erethizon dorsatum bruneri* (Anderson, Bull. Nat. Mus. Canada, 102:173, 24 January 1947). Original label indicates specimen was found by collector in the trees on his farm and sent to M. H. Swenk on 30 September; kept in captivity until 19 October, when killed.

CETACEA

Phocaenidae

Phocoena sinus Norris and McFarland

J. Mamm., 39(1):24, 20 February 1958.

Type. — Adult, sex unknown, skull only, MVZ 120933; from northeast shore of Punta San Felipe, Baja California, Mexico; picked up in sand by K. S. Norris on 18 March 1950, no original number; no measurements.

Remarks. — Angular process of right dentary missing.

CARNIVORA

Canidae

Canis latrans dickeyi Nelson

Proc. Biol. Soc. Washington, 45:224, 26 November 1932.

Type. — First year male, study skin and skull, MVZ 132043; from 2 miles west of Rio Goascoran [near Cerro Mogote], 13° 30' N latitude, Dist. La Union, El Salvador; collected by G. D. Stirton on 29 December 1926, original number (collection of Donald R. Dickey) 12260; measurements: 1286-380-250-80.

Canis latrans incolatus Hall

Univ. California Publ. Zool., 40(9):369, 5 November 1934.

Type. — Adult female, flat skin and skull, MVZ 43898; from Isaacs Lake, 3000 feet altitude [Bowron Lake region, Barkerville District], British Columbia; collected by T. T. and E. B. McCabe on 23 October 1928, original number 201; measurements: 1099-255-181.

Canis occidentalis crassodon Hall

Univ. California Publ. Zool., 38(12):420, 8 November 1932.

Type.- Adult male, skull only, MVZ 12456; from Thasis Canal, Nootka Sound, Vancouver Island, British Columbia; collected by Carl Leiner during winter of 1909-1910, obtained by H. S. Swarth on 26 July 1910, original number (H. S. Swarth) 8419.

Remarks. — *Canis lupus crassodon* (Goldman, J. Mamm., 18:45, 11 February 1937).

Vulpes macrotis zinseri Benson

Proc. Biol. Soc. Washington, 51:21, 18 February 1938.

Type. — Adult male, flat skin and skull plus baculum, MVZ 76292; from 1/4 mile southeast of San Antonio de Jaral [El Jarral of most maps, approx. 25° 40′ N latitude, 101° 25′ W longitude], 4400 feet altitude, Coahuila, Mexico; collected by G. Rodriguez on 15 June 1937, original number (W. B. Richardson) 2568; measurements: 822-325-123-80-88, 2098.2 g.

Urocyon californicus sequoiensis Dixon

Univ. California Publ. Zool., 5(7):303, 12 February 1910.

Type. — Adult female, flat skin and skull plus atlas and hyoid bones, MVZ 8978; from Lagunitas, Marin County, California; collected by J. Dixon on 1 January 1910, original number 455; measurements: 890-320-127.

Remarks. — A synonym of *Urocyon cinereoargenteus townsendi* according to Grinnell (Univ. California Publ. Zool., 40:110, 26 September 1933).

Ursidae

Ursus alexandrae Merriam

Proc. Biol. Soc. Washington, 27:174, 13 August 1914.

Type. — Adult male, skin with complete skeleton, MVZ 4752; from Kasilof Lake [Kenai Peninsula], Alaska; collected by Andrew Berg in September 1906, original number (A. M. Alexander) 218; no measurements.

Remarks. — A synonym of *Ursus arctos horribilis* according to Rausch (Arctic, 6:105, July 1953).

Ursus americanus pugnax Swarth

Univ. California Publ. Zool., 7(2):141, 12 January 1911.

Type. — Adult male, skin and skull, MVZ 8332; from Rocky Bay, Dall Island, Alaska; collected by A.E. Hasselborg on 31 May 1909, original number 15; no measurements.

Remarks. — Left lower incisors and canine broken off.

Ursus americanus vancouveri Hall

Univ. California Publ. Zool., 30(10):231, 2 March 1928.

Type.— Adult male, skin with complete skeleton, MVZ 12461; from King Solomon Basin, Vancouver Island, 20 miles south of Alberni, British Columbia, collected by E. Despard on 16 July 1910, original number 38; measurements: 1540-x-252-107.

Procyonidae

Bassariscus astutus octavus Hall

Univ. California Publ. Zool., 30(3):39, 8 September 1926.

Type.— Old adult male, skin with complete skeleton, MVZ 27860; from San Luis Rey River, 1700 feet altitude; near Escondido, San Diego County, California; obtained by James B. Dixon on 6 March 1918 and sent alive to Joseph Dixon; prepared by him on 27 March 1918, original number (J. Dixon) 6624; measurements: 775-400-73-39, 1026 g.

Remarks.— Original description erroneously indicates specimen collected in 1925.

Procyon lotor dickeyi Nelson and Goldman

Proc. Biol. Soc. Washington, 44:18, 21 February 1931.

Type.— Adult male, skin and skull, MVZ 132091; from Barra de Santiago, Dept. Ahuachapan, El Salvador; collected by G. D. Stirton on 14 April 1927, original number (collection of Donald R. Dickey) 12796; measurements: 840-310-115-46.

Remarks.— Molar teeth very worn. Original label indicates specimen was shot in tree on swamp island near water hole; stomach=crabs.

Mustelidae

Lutreola vison nesolestes Heller

Univ. California Publ. Zool., 5(2):259, 18 February 1909.

Type.— Adult male, skin and skull, MVZ 201; from Windfall Harbor [Admiralty Island], Alaska; collected by J. Shaga on 28 April 1907, original number (A. M. Alexander) 14; measurements: 615-182-80, 4-3/4 lbs.

Remarks.— *Mustela vison nesolestes* (Miller, Bull, U. S. Nat. Mus., 79:102, 31 December 1912). The original description indicates the specimen was collected by A. M. Alexander, but the original label has J. Shaga as collector. Miss Alexander's field catalog includes this specimen, but also indicates that a river otter (no. 26) taken the same day was collected by J. Shaga. I am inclined to believe this holotype was secured by J. Shaga and turned over to Miss Alexander for preparation.

Mustela cicognanii anguinae Hall

Univ. California Publ. Zool., 38(12):417, 8 November 1932.

Type.— Adult male, complete skeleton only, MVZ 12482; from French Creek, Vancouver Island, British Columbia; found as a desiccated carcass by W. H. Lee on 1 May 1910, no original number; measurements (from dried carcass by E. R. Hall): 235-70-34-14.

Remarks.— Mustela erminea anguinae (Hall, J. Mamm., 26:79, 27 February 1945).

Mustela erminea initis Hall

Proc. Biol. Soc. Washington, 57:37, 28 June 1944.

Type.— Adult male, skin and skull, MVZ 289; from Saook Bay [Baranof Island], Alaska; collected by A. E. Hasselborg on 9 October 1911, original number 4 (A. M. Alexander original number 81); measurements: 330-95-45.

Mustela erminea invicta Hall

J. Mamm., 26(1):75, 23 February 1945.

Type.— Subadult male, skin and skull, MVZ 101122; from Benewah [east of Tekoa, Whitman County, Washington], Benewah County, Idaho; collected by William T. Shaw on 24 October 1926, no original number; measurements: 280-75-34, 81 g.

Mustela erminea salva Hall

Proc. Biol. Soc. Washington, 57:35, 28 June 1944.

Type.— Adult male, skull only, MVZ 74641; from Mole Harbor, Admiralty Island, Alaska; collected by A. E. Hasselborg on 27 December 1936, original number (E. R. Hall) 5450.

Mustela erminea seclusa Hall

Proc. Biol. Soc. Washington, 57:39, 28 June 1944.

Type.— Adult male, skull only, MVZ 31232; from Port Santa Cruz, Suemez Island, Alaska; collected by G. Willett on 24 March 1920, no original number; no measurements.

Mustela frenata altifrontalis Hall

Carnegie Inst. Washington Publ. No. 473:94, 20 November 1936.

Type.— Adult male, skin and skull, MVZ 42093; from Tillamook [Tillamook County], Oregon; collected by Alex Walker on 10 July 1928, original number 717; measurements: 438-167-53.

Mustela frenata inyoensis Hall

Carnegie Inst. Washington Publ. No. 473:99, 20 November 1936.

Type.— Adult male, skin with complete skeleton, MVZ 25907; from Carl Walter's Ranch [at the site of old Fort Independence, 3900 feet altitude], 2 miles north of Independence, Inyo County, California; collected by A. C. Shelton on 26 June 1917, original number 3143; measurements: 423-170-42-16.

Mustela frenata latirostra Hall

Carnegie Inst. Washington Publ. No. 473:96, 20 November 1936.

Type.— Adult male, skin and skull, MVZ 3257; from San Diego [San Diego County], California; collected by Frank X. Holzner on 20 May 1907; no original number; measurements: 435-142-42-16 (crown).

Mustela frenata nevadensis Hall

Carnegie Inst. Washington Publ. No. 473:91, 20 November 1936.

Type.— Adult female, skin with complete skeleton, MVZ 41503; from 3 miles east of Baker, White Pine County, Nevada; collected by E. R. Hall and W. C. Russell on 30 May 1929, original number (E. R. Hall) 2674; measurements: 333-123-37-26.3 123.6 g.

Mustela frenata nigriauris Hall

Carnegie Inst. Washington Publ. No. 473:95, 20 November 1936.

Type.— Adult male, skin with complete skeleton, MVZ 70210; from 2-1/2 miles east of Santa Cruz, Santa Cruz County, California; collected by C. H. Russell on 11 March 1936, original number (W. C. Russell) 4225; measurements: 410-143-45-27.

Mustela frenata pulchra Hall

Carnegie Inst. Washington Publ. No. 473:98, 20 November 1936.

Type.— Adult male, skin with complete skeleton, MVZ 16668; from Buttonwillow, Kern County, California; collected by J. Grinnell on 30 April 1912, prepared by F. H. Holden, original number (J. Grinnell) 1953; measurements: 460-184-49-27.

Mustela noveboracensis arthuri Hall

Proc. Biol. Soc. Washington, 40:193, 2 December 1927.

Type.— Subadult male, skin and skull, MVZ 37515; from Remy, St. James Parish, Louisiana; collected by S. C. Arthur on 12 December 1926, original number 3; measurements (inches): 15-1/2 - 4-1/2 - 1-3/4.

Remarks.— *Mustela frenata arthuri* (Hall, Carnegie Inst. Washington Publ. No. 473:105, 20 November 1936). In 1958 A. C. Ziegler noted that a letter from S. C. Arthur dated 28 December 1926 (since misplaced) gives date of collection as "on or about December 15, 1926," the specimen label gives the date as about 12,26,1926. The "12" has been corrected and may read "15."

Mustela vison aestuarina Grinnell

Proc. Biol. Soc. Washington, 29:213, 22 September 1916.

Type.— Adult male, flat skin with complete skeleton, MVZ 23660; from Grizzly Island, Solano County, California; trapped for Miss Annie M. Alexander by A. H. Luscomb and received at MVZ on 30 November 1915, prepared by F. H. Holden, original number (F. H. Holden) 925; measurements: 594-183-74, 2-1/2 lbs.

Mustela vison evagor Hall

Univ. California Publ. Zool., 38(12):418, 8 November 1932.

Type. – Adult male, skin with complete skeleton, MVZ 12479; from [Little Qualicum River, eight to nine miles west of Parksville], Vancouver Island, British Columbia; collected by E. Despard on 11 May 1910, original number 4; measurements: 596-190-80.

Martes caurina humboldtensis Grinnell and Dixon

Univ. California Publ. Zool., 21(16):411, 17 March 1926.

Type. – Subadult male, skull only, MVZ 19158; from [ridge about] 5 miles northeast of Cuddeback [=Carlotta], Humboldt County, California; obtained from local trapper by H. E. Wilder in February 1913, original number (H. E. Wilder) 1368; no measurements.

Remarks. – *Martes americana humboldtensis* (Wright, J. Mamm., 34:85, 9 February 1953).

Martes caurina sierrae Grinnell and Storer

Univ. California Publ. Zool., 17(1):2, 23 August 1916.

Type. – Adult male, skin with complete skeleton, MVZ 22112; from head of Lyell Canyon, 9800 feet altitude, Yosemite National Park, Tuolumne County, California; collected by C. D. Holliger on 24 July 1915, original number 562; measurements: 609-190-79-35, 950 g.

Remarks. – *Martes americana sierrae* (Wright, J. Mamm., 34:85, 9 February 1953).

Martes caurina vancouverensis Grinnell and Dixon

Univ. California Publ. Zool., 21(16):414, 17 March 1926.

Type. – Adult male, skin and skull, MVZ 12474; from Golden Eagle Mine, 20 miles south of Alberni, Vancouver Island, British Columbia; collected by E. Despard on 20 July 1910, original number 41; measurements: 590-200-90-40.

Remarks. – *Martes americana vancouverensis* (Wright, J. Mamm., 34:85, 9 February 1953).

Spilogale phenax microrhina Hall

J.Mamm., 7(1):53, 15 February 1926.

Type. – Adult male, skin and skull, MVZ 3215; from Julian, San Diego County, California; collected by Frank Stephens on 29 July 1908, original number 1308; measurements: 405-157-47-20.

Remarks. – A synonym of *Spilogale putorius phenax* according to Van Gelder, (Bull. Amer. Mus. Nat. Hist., 117(5):335, 15 June 1959). Mead (J. Mamm., 49:373, 20 August 1968) suggests further study may assign this subspecies to *Spilogale gracilis*.

Lutra canadensis brevipilosus Grinnell

Univ. California Publ. Zool., 12(8):306, 31 October 1914.

Type. — Adult female, flat skin with complete skeleton, MVZ 20775; from Grizzly Island [Suisun Bay], Solano County, California; secured from a local trapper (probably A. H. Luscomb) by Miss Annie M. Alexander, and presented by her to the museum on 26 January 1914; prepared by F. H. Holden, no original number; measurements: 1158-447-123.5-23.8, 16 lbs, 10 oz.

Lutra canadensis interior Swenk

Univ. Stud., Lincoln, Nebraska, 18(1):2, 15 May 1920.

Type. — Old adult male, study skin and skull, MVZ 28728; from Lincoln Creek, west of Seward, Seward County, Nebraska; found dead in creek by George and R. Anderson on 4 June 1916, received by Myron H. Swenk through Henry Heumann, no original number; measurements: (from made-up skin): 1270-488-120-16.5.

ARTIODACTYLA

Cervidae

Odocoileus hemionus fuliginatus Cowan

J. Mamm., 14(4):326, 13 November 1933.

Type. — Adult male, flat skin and skull, MVZ 39918; from Barona Ranch, 30 miles east of San Diego, San Diego County, California; seized by deputies Webb Toms and Glidden on 8 October 1928 and received at MVZ on 24 October 1928, no original number, no measurements.

Remarks. — California Fish and Game Commission label list James D. Bobbitt as "owner or shipper."

Odocoileus hemionus inyoensis Cowan

Proc. Biol. Soc. Washington, 46:69, 27 April 1933.

Type. — Adult male, skin with complete skeleton, MVZ 16363; from Kid Mountain, 11000 feet altitude, 10 miles west of Big Pine, Inyo County, California; collected by H. A. Carr on 15 October 1911, original number 656; measurements: 1740-180-485-210, approx. 190 lbs.

Remarks. — Premaxillary bones missing.

Bovidae

Ovis cervina sierrae Grinnell

Univ. California Publ. Zool., 10(5):144, 9 May 1912.

Type. — Male (age about 5 years), skin with complete skeleton, MVZ 16360; from the east slope of Mount Baxter, 11000 feet altitude, Sierra Nevada, Inyo County, California; collected by H. A. Carr on 20 October 1911, original number

659; measurements: 1570-100-420-110, 200 lbs.

Remarks.— A synonym of *Ovis canadensis californiana* according to Cowan (Amer. Midland Natur., 24:554, November 1940). Note on skin label indicates the head region was in poor condition and repaired on 17 October 1921.

GEOGRAPHIC ORIGINS OF TYPE SPECIMENS

(arranged alphabetically within each geographic division)

ARGENTINA
    Phyllotis caprinus Pearson

CANADA
  British Columbia
    Canis latrans incolatus Hall
    Canis occidentalis crassodon Hall
    Castor canadensis sagittatus Benson
    Marmota vancouverensis Swarth
    Martes caurina vancouverensis Grinnell and Dixon
    Microtus pennsylvanicus funebris Dale
    Microtus pennsylvanicus rubidus Dale
    Microtus townsendii cummingi Hall
    Mustela cicognanni anguinae Hall
    Mustela vison evagor Hall
    Peromyscus maniculatus angustus Hall
    Peromyscus maniculatus georgiensis Hall
    Sorex obscurus mixtus Hall
    Ursus americanus vancouveri Hall

  Nova Scotia
    Blarina brevicauda pallida Smith
    Clethrionomys gapperi pallescens Hall and Cockrum
    Condylura cristata nigra Smith
    Peromyscus leucopus caudatus Smith
    Sorex arcticus maritimensis Smith

EL SALVADOR
    Canis latrans dickeyi Nelson
    Nyctomys sumichrasti florencei Goldman
    Procyon lotor dickeyi Nelson and Goldman
    Reithrodontomys mexicanus orinus Hooper

MEXICO

Baja California

Microtus californicus perplexabilis Grinnell
Myotis yumanensis lambi Benson
Neotoma fuscipes martirensis Orr
Neotoma lepida egressa Orr
Perognathus formosus cinerascens Nelson and Goldman
Perognathus spinatus lambi Benson
Peromycscus eremicus cinereus Hall
Phocoena sinus Norris and McFarland
Thomomys bottae lucidus Hall
Thomomys bottae siccovallis Huey

Chihuahua

Eutamias dorsalis nidoensis Lidicker
Reithrodontomys fulvescens canus Benson

Coahuila

Vulpes macrotis zinseri Benson

Michoacán

Liomys irroratus acutus Hall and Villa
Microtus mexicanus fundatus Hall
Sigmodon hispidus atratus Hall
Thomomys umbrinus pullus Hall and Villa

Sonora

Dipodomys spectabilis intermedius Nader
Eumops sonoriensis Benson
Peromyscus crinitus delgadilli Benson
Peromyscus crinitus rupicolus Benson
Peromyscus crinitus scopulorum Benson
Reithrodontomys burti Benson
Thomomys bottae basilicae Benson and Tillotson
Thomomys bottae estanciae Benson and Tillotson

PARAGUAY

Graomys pearsoni Myers

UNITED STATES

Alaska

Castor canadensis belugae Taylor
Castor canadensis phaeus Heller
Citellus lyratus Hall and Gilmore
Clethrionomys albiventer Hall and Gilmore
Clethrionomys dawsoni glacialis Orr
Evotomys dawsoni insularis Heller
Evotomys phaeus Swarth

Lutreola vison nesolestes Heller
Marmota caligata broweri Hall and Gilmore
Marmota ochracea Swarth
Marmota vigilis Heller
Microtus admiraltiae Heller
Microtus coronarius Swarth
Microtus innuitus punukensis Hall and Gilmore
Microtus mordax littoralis Swarth
Mustela erminea initis Hall
Mustela erminea salva Hall
Mustela erminea seclusa Hall
Peromyscus sitkensis oceanicus Cowan
Sciurus hudsonicus picatus Swarth
Sorex jacksoni Hall and Gilmore
Sorex obscurus malitiosus Jackson
Ursus alexandrae Merriam
Ursus americanus pugnax Swarth

Arizona

Citellus tereticaudus arizonae Grinnell
Dipodomys merriami vulcani Benson
Neotoma lepida flava Benson
Perognathus amplus ammodytes Benson
Perognathus amplus cineris Benson
Perognathus intermedius crinitus Benson
Perognathus intermedius umbrosus Benson
Perognathus parvus trumbullensis Benson
Sigmodon ochrognathus montanus Benson
Sorex melanogenys Hall
Thomomys bottae nasutus Hall
Thomomys bottae trumbullensis Hall and Davis
Thomomys chrysonotus Grinnell
Thomomys harquahalae Grinnell and Hill
Thomomys perpallidus depauperatus Grinnell and Hill

California

Ammospermophilus nelsoni amplus Taylor
Aplodontia chryseola Kellogg
Aplodontia humboltiana Taylor
Aplodontia nigra Taylor
Bassariscus astutus octavus Hall
Callospermophilus chrysodeirus perpallidus Grinnell
Castor subauratus Taylor
Corynorhinus macrotis intermedius Grinnell
Dipodomys agilis fuscus Boulware
Dipodomys berkeleyensis Grinnell
Dipodomys californicus eximius Grinnell

Dipodomys californicus trinitatis Kellogg
Dipodomys deserti aquilus Nader
Dipodomys heermanni arenae Boulware
Dipodomys heermanni saxatilis Grinnell and Linsdale
Dipodomys jolonensis Grinnell
Dipodomys merriami brevinasus Grinnell
Dipodomys merriami collinus Lidicker
Dipodomys panamintinus caudatus Hall
Dipodomys sanctiluciae Grinnell
Eutamias amoenus monoensis Grinnell and Storer
Eutamias merriami kernensis Grinnell and Storer
Eutamias merriami mariposae Grinnell and Storer
Eutamias minimus scrutator Hall and Hatfield
Eutamias panamintinus acrus Johnson
Eutamias sonomae Grinnell
Glaucomys sabrinus flaviventris Howell
Lepus washingtonii tahoensis Orr
Lutra canadensis brevipilosus Grinnell
Marmota flaviventris fortirostris Grinnell
Martes caurina humboldtensis Grinnell and Dixon
Martes caurina sierrae Grinnell and Storer
Microdipodops polionotus Grinnell
Microtus californicus aestuarinus Kellogg
Microtus californicus eximius Kellogg
Microtus californicus halophilus von Bloeker
Microtus californicus kernensis Kellogg
Microtus californicus mariposae Kellogg
Microtus californicus mohavensis Kellogg
Microtus californicus paludicola Hatfield
Microtus californicus sanctidiegi Kellogg
Microtus californicus sanpabloensis Thaeler
Microtus montanus yosemite Grinnell
Microtus mordax sierrae Kellogg
Mustela frenata inyoensis Hall
Mustela frenata latirostra Hall
Mustela frenata nigriauris Hall
Mustela frenata pulchra Hall
Mustela vison aestuarina Grinnell
Myotis californicus quercinus Grinnell
Myotis californicus stephensi Dalquest
Myotis lucifugus relictus Harris
Myotis ruddi Silliman and von Bloeker
Myotis yumanensis altipetens Grinnell
Myotis yumanensis oxalis Dalquest
Myotis yumanensis sociabilis Grinnell
Neotoma cinerea alticola Hooper
Neotoma cinerea pulla Hooper

Neotoma fuscipes bullatior Hooper
Neotoma fuscipes luciana Hooper
Neotoma fuscipes perplexa Hooper
Neotoma fuscipes riparia Hooper
Neotoma lepida grinnelli Hall
Neotoma lepida petricola von Bloeker
Ochotona albatus Grinnell
Ochotona schisticeps muiri Grinnell and Storer
Ochotona schisticeps sheltoni Grinnell
Ochotona taylori Grinnell
Odocoileus hemionus fuliginatus Cowan
Odocoileus hemionus inyoensis Cowan
Ovis cervina sierrae Grinnell
Perodipus dixoni Grinnell
Perodipus elephantinus Grinnell
Perodipus leucogenys Grinnell
Perodipus mohavensis Grinnell
Perodipus monoensis Grinnell
Perodipus swarthi Grinnell
Perognathus californicus bensoni von Bloeker
Perognathus californicus bernardinus Benson
Perognathus californicus marinensis von Bloeker
Perognathus inornatus sillimani von Bloeker
Perognathus longimembris cantwelli von Bloeker
Perognathus longimembris neglectus Taylor
Perognathus longimembris psammophilus von Bloeker
Perognathus longimembris tularensis Richardson
Perognathus xanthonotus Grinnell
Peromyscus californicus benitoensis Grinnell and Orr
Peromyscus californicus mariposae Grinnell and Orr
Peromyscus truei chlorus Hoffmeister
Peromyscus truei sequoiensis Hoffmeister
Reithrodontomys halicoetes Dixon
Reithrodontomys megalotis distichlis von Bloeker
Reithrodontomys megalotis santacruzae Pearson
Reithrodontomys raviventris Dixon
Scapanus latimanus campi Grinnell and Storer
Scapanus latimanus caurinus Palmer
Scapanus latimanus grinnelli Jackson
Scapanus latimanus insularis Palmer
Scapanus latimanus monoensis Grinnell
Scapanus latimanus occultus Grinnell and Swarth
Scapanus latimanus parvus Palmer
Sorex halicoetes Grinnell
Sorex montereyensis mariposae Grinnell
Sorex ornatus relictus Grinnell
Sorex ornatus salarius von Bloeker

Sorex ornatus salicornicus von Bloeker
Sorex pacificus sonomae Jackson
Sorex sinuosus Grinnell
Spilogale phenax microrhina Hall
Sylvilagus bachmani macrorhinus Orr
Sylvilagus bachmani mariposae Grinnell and Storer
Sylvilagus bachmani riparius Orr
Sylvilagus bachmani tehamae Orr
Thomomys albatus Grinnell
Thomomys bottae acrirostratus Grinnell
Thomomys bottae agricolaris Grinnell
Thomomys bottae crassus Chattin
Thomomys bottae ingens Grinnell
Thomomys bottae piutensis Grinnell and Hill
Thomomys bottae rupestris Chattin
Thomomys bottae saxatilis Grinnell
Thomomys bottae silvifugus Grinnell
Thomomys diaboli Grinnell
Thomomys infrapallidus Grinnell
Thomomys jacinteus Grinnell and Swarth
Thomomys melanotis Grinnell
Thomomys monticola premaxillaris Grinnell
Thomomys nigricans puertae Grinnell
Thomomys perpallidus amargosae Grinnell
Thomomys perpallidus mohavensis Grinnell
Thomomys perpallidus riparius Grinnell and Hill
Thomomys providentialis Grinnell
Thomomys relictus Grinnell
Urocyon californicus sequoiensis Dixon
Zapus trinotatus eureka Howell

Colorado

Castor canadensis concisor Warren and Hall
Microtus montanus fusus Hall
Spermophilus tridecemlineatus blanca Armstrong

Idaho

Castor canadensis taylori Davis
Citellus elegans aureus Davs
Dipodomys microps idahoensis Hall and Dale
Lepus californicus depressus Hall and Whitlow
Mustela erminea invicta Hall
Ochotona princeps clamosa Hall and Bowlus
Peromyscus maniculatus serratus Davis
Thomomys townsendii owyhensis Davis
Thomomys townsendii similis Davis
Zapus princeps idahoensis Davis

Louisiana
> *Mustela noveboracensis arthuri* Hall

Montana
> *Dipodomys ordii terrosus* Hoffmeister
> *Thomomys talpoides confinis* Davis

Nebraska
> *Erethizon epixanthum bruneri* Swenk
> *Geomys bursarius majusculus* Swenk
> *Geomys lutescens levisagittalis* Swenk
> *Geomys lutescens vinaceus* Swenk
> *Lutra canadensis interior* Swenk
> *Perognathus flavescens olivaceogriseus* Swenk
> *Pteromys volans nebrascensis* Swenk
> *Thomomys talpoides cheyennensis* Swenk
> *Thomomys talpoides pierreicolus* Swenk

Nevada
> *Callospermophilus trepidus* Taylor
> *Citellus beldingi crebrus* Hall
> *Dipodomys microps centralis* Hall and Dale
> *Dipodomys microps occidentalis* Hall and Dale
> *Dipodomys ordii fetosus* Durrant and Hall
> *Dipodomys ordii inaquosus* Hall
> *Eutamias amoenus celeris* Hall and Johnson
> *Microdipodops megacephalus ambiguus* Hall
> *Microdipodops megacephalus medius* Hall
> *Microdipodops megacephalus nasutus* Hall
> *Microdipodops megacephalus nexus* Hall
> *Microdipodops megacephalus sabulonis* Hall
> *Microdipodops pallidus albiventer* Hall and Durrant
> *Microdipodops pallidus ammophilus* Hall
> *Microdipodops pallidus purus* Hall
> *Microdipodops pallidus ruficollaris* Hall
> *Microtus (Lagurus) intermedius* Taylor
> *Microtus montanus fucosus* Hall
> *Microtus montanus micropus* Hall
> *Microtus montanus undosus* Hall
> *Microtus mordax latus* Hall
> *Mustela frenata nevadensis* Hall
> *Neotoma nevadensis* Taylor
> *Ochotona princeps tutelata* Hall
> *Perognathus formosus incolatus* Hall
> *Perognathus formosus melanurus* Hall
> *Perognathus longimembris gulosus* Hall
> *Peromyscus truei nevadensis* Hall and Hoffmeister

Thomomys bottae abstrusus Hall and Davis
Thomomys bottae brevidens Hall
Thomomys bottae cinereus Hall
Thomomys bottae concisor Hall and Davis
Thomomys bottae curtatus Hall
Thomomys bottae depressus Hall
Thomomys bottae fumosus Hall
Thomomys bottae lacrymalis Hall
Thomomys bottae latus Hall and Davis
Thomomys bottae lucrificus Hall and Durham
Thomomys bottae nanus Hall
Thomomys bottae vescus Hall and Davis
Thomomys falcifer Grinnell
Thomomys perpallidus centralis Hall
Thomomys solitarius Grinnell
Thomomys townsendii bachmani Davis
Thomomys townsendii elkoensis Davis
Zapus princeps curtatus Hall
Zapus princeps palatinus Hall

New Jersey

Sorex cinereus nigriculus Green

New Mexico

Citellus grammurus tularosae Benson
Geomys arenarius brevirostris Hall
Perognathus intermedius rupestris Benson
Peromyscus nasutus griseus Benson
Thomomys baileyi tularosae Hall
Thomomys bottae connectens Hall
Thomomys bottae ruidosae Hall

Oregon

Castor canadensis idoneus Jewett and Hall
Lepus bairdii oregonus Orr
Mustela frenata altifrontalis Hall
Thomomys bottae detumidus Grinnell
Thomomys quadratus wallowa Hall and Orr

Texas

Dipodomys ordii attenuatus Bryant

Utah

Glaucomys sabrinus lucifugus Hall
Microdipodops megacephalus paululus Hall and Durrant
Microtus mexicanus navaho Benson
Microtus montanus amosus Hall and Hayward
Microtus montanus nexus Hall and Hayward
Neotoma albigula brevicauda Durrant

Ochotona princeps utahensis Hall and Hayward
Perognathus longimembris arcus Benson
Thomomys perpallidus albicaudatus Hall
Thomomys perpallidus aureiventris Hall
Thomomys quadratus gracilis Durrant
Zapus princeps cinereus Hall

Washington

Aplodontia rufa grisea Taylor
Microtus montanus kincaidi Dalquest
Microtus mordax halli Hayman and Holt
Microtus townsendii pugeti Dalquest
Myotis evotis pacificus Dalquest
Neurotrichus gibbsii minor Dalquest and Burgner
Thomomys talpoides aequalidens Dalquest
Thomomys talpoides devexus Hall and Dalquest
Thomomys talpoides immunis Hall and Dalquest
Thomomys talpoides yakimensis Hall and Dalquest
Scapanus orarius yakimensis Dalquest and Scheffer

Wyoming

Dipodomys ordii priscus Hoffmeister

Literature Cited

GRINNELL, J.
    1933.   Review of the Recent mammal fauna of California. Univ. California Publ. Zool., 40(2):71-234.

MILLER, G. S., JR., and R. KELLOGG
    1955.   List of North American Recent mammals. Bull. U. S. Nat. Mus., No. 205:1-954.

SIMPSON, G. G.
    1945.   The principles of classification and a classification of mammals. Bull. Amer. Mus. Nat. Hist., 85:1-350.

Forsyth Library
F.H.S.U.

Recent Volumes in
University of California Publications in Zoology

Volume 109. Allen, Merlin and Ella Mar Noffsinger. A Revision of the Marine Nematodes of the Superfamily Draconematoidea Filipjev, 1918 (Nematoda: Draconematina). ISBN 0-520-09583-9.

Volume 112. Johnson, Ned K. Character Variation of Evolution of Sibling Species in the Empidonax Difficilis-Flavescens Complex (Aves: Tyrannidae). ISBN 0-520-09599-5.

Volume 113. Howell, Thomas R. Breeding Biology of the Egyptian Plover Pluvianus Argyptius (Aves: Glareolidae). ISBN 0-520-03804-5.

ISBN 0-520-09622-3